OPERATIONAL 1 FORECASTING, WAR RESPONSE FOR MULTI-SCALE FLOOD RISKS IN DEVELOPING CITIES

OPERATIONAL FLOOD FORECASTING, WARNING AND RESPONSE FOR MULTI-SCALE FLOOD RISKS IN DEVELOPING CITIES

DISSERTATION

Submitted in fullfillment of the requirements of
the Board for Doctorates of Delft University of Technology
and
of the Academic Board of the UNESCO-IHE
Institute for Water Education
for
the Degree of DOCTOR
to be defended in public
on Monday, 13 June 2016, at 15:00 hours
in Delft, the Netherlands

by

María Carolina ROGELIS PRADA

Civil Engineer, National University of Colombia
Master of Engineering, Los Andes University
Master of Science in Hydraulic Engineering, UNESCO-IHE
born in Bogotá, Colombia

This dissertation has been approved by the
promotor: Prof.dr.ir A. E. Mynett and
copromotor: Dr.ir. M.G.F. Werner

CRC Press/Balkema is an imprint of the Taylor & Francis Group, an informa business

Published by:
CRC Press/Balkema
PO Box 11320, 2301 EH Leiden, the Netherlands
Pub.NL@taylorandfrancis.com
www.crcpress.com – www.taylorandfrancis.com
ISBN 978-1-138-03003-9

To my husband, Juan, for always supporting, helping, and standing by me.

Summary

Flood early warning systems are recognized as one of the most effective flood risk management instruments when correctly embedded in comprehensive flood risk management strategies and policies. Many efforts around the world are being put in place to advance the components that determine the effectiveness of a flood early warning system. The aim of this research is to contribute to the understanding of the risk knowledge and forecasting components of flood early warning in the particular environment of tropical high mountains in developing cities. These represent a challenge taking into account the persistent lack of data, limited resources and often complex climatic, hydrologic and hydraulic conditions. The contributions of this research are intended to advance the knowledge required for design and operation of flood early warning in data-scarce watersheds from a hydrological perspective, without neglecting the crosscutting nature of flood early warning in the flood risk management process.

Risk knowledge provides the framework for the operation of flood early warning systems. In this research, a regional method for assessing flash flood susceptibility and for identifying debris flow predisposition at the watershed scale is proposed. The method is based on an index composed of a morphometric indicator and a land cover indicator, which is applied in 106 peri-urban mountainous watersheds in Bogotá, Colombia. The susceptibility indicator is obtained from readily available information common to most peri-urban mountainous areas and can be used to prioritise watersheds that can subsequently be subjected to a more detailed hazard analysis. The indicator is useful in the identification of flood type, which is a crucial step in flood risk assessment especially in mountainous environments; and it can be used as input for prioritization of flood risk management strategies at regional level and for the prioritization and identification of detailed flood hazard analysis. The indicator is regional in scope and therefore it is not intended to constitute a detailed assessment but to highlight watersheds that could potentially be more susceptible to damaging floods than others in the same region.

The analysis of risk requires the assessment of both hazard and vulnerability. An indication of hazard was obtained from the flash flood susceptibility analysis and then, vulnerability at watershed scale was obtained. Vulnerability was assessed on the basis of a principal component analysis carried out with variables recognised in literature to contribute to vulnerability. Once the vulnerability indicator was obtained, this was combined with the susceptibility indicator, thus providing an index that allows the watersheds to be prioritised in support of flood risk management at regional level. The complex interaction between vulnerability and hazard is evidenced in the case study. Environmental degradation in vulnerable watersheds shows the influence that vulnerability exerts on hazard and vice versa, thus establishing a cycle that builds up risk conditions.

Once priority watersheds for flood risk management measures have been identified based on risk analyses, the research follows the modelling steps for flood forecasting development. As first step, input precipitation is addressed in the environment of complex topography commonly found in mountainous tropical areas. The difference in performance of interpolation techniques (Ordinary Kriging and Kriging with external Drift) is assessed in order to propose a real time operational procedure to obtain rainfall fields from gauged data. The performance of pooled variograms and the added value of secondary variables in the interpolation procedure were studied. The results showed that interpolators using pooled variograms provide a performance comparable to when the interpolators were applied to the storms individually, showing that they can be used successfully for interpolation in real-time operation in the study area. Furthermore, the analysis identified limitations in the use of Kriging with External Drift. Only when the correlation between the secondary variables and precipitation is higher than the percentage of variability explained found in Ordinary Kriging, then Kriging with External Drift provided a consistent improvement.

Models are the heart of flood forecasting systems. As such, the choice among possible hydrological models constitutes a key issue. This is a challenge in high tropical mountain areas, particularly in páramos (tropical high mountain ecosystems). These have been considered sacred places by the indigenous population of Latin America and are recognized as areas with an immense natural value. Furthermore páramos are the source of water for many important cities in the Andes. In contrast to their great importance, the knowledge about their hydrologic process is still very limited. In this research a distributed model (TETIS), a semi-distributed model (TOPMODEL) and a lumped model (HEC HMS soil moisture accounting) were used to simulate the discharges of a tropical high mountain basin with a páramo upper basin. Performance analysis and diagnostics were carried out using the signatures of the flow duration curve and through analysis of the model fluxes in order to identify the most appropriate model for the study area for flood early warning. The impact of grid size was explored in the distributed and semi-distributed models in order to provide insight into the criteria to be used for forecasting modelling. The sensitivity of the models to variation in the precipitation input was analysed by forcing the models with a rainfall ensemble obtained from Gaussian simulation. The resulting discharge ensembles of each model were compared in order to identify differences among models structures. The results show that TOPMODEL is the most realistic model of the three tested, albeit showing the larger discharge ensemble spread.

Numerical Weather Prediction (NWP) models are fundamental to extend lead-times beyond the concentration time of a watershed. NWPs are increasingly used in flood forecasting centres around the world. In this research, the WRF model under the settings currently used by the National Meteorological Agency to issue weather forecasts in Bogotá (Colombia) was used to explore its added value for flood early warning in a páramo area. Forecasts generated under four strategies were used to drive the hydrological model constructed for the study area: a) Zero rainfall forecasts; b) Raw forecasts from the WRF; c) deterministic bias corrected WRF forecasts; d) and precipitation forecast ensembles obtained from the WRF model. In order to assess the value of the streamflow forecasts obtained from driving the hydrologic model with the WRF forecasts, a reference forecast equal to the obtained from forecast precipitation equal to zero was used. Results show that the streamflow forecasts obtained from a hydrological model driven by post-processed WRF precipitation add value to the flood early warning system when compared to zero precipitation forecasts. Despite the fact that the added value of the WRF model forecasts is modest, this shows promise for increasing forecast skill in areas of high meteorological and topographic complexity and the possibility of improvement.

Samenvatting

Waarschuwingssystemen tegen overstromingen worden gezien als een van de meest effectieve manieren om overstromingsrisico's te beperken. Voorwaarde daarbij is dat deze systemen op de juiste manier zijn ingebed in een zorgvuldig voorbereide strategie van beleidsmaatregelen. Over de hele wereld wordt op verschillende manieren gewerkt aan de vele onderdelen die de effectiviteit van dergelijke systemen bepalen. Het doel van het onderzoek in dit proefschrift is om bij te dragen aan het verbeteren van de kennis op het gebied van risico analyse en de voorspelbaarheid van overstromingen in stedelijke gebieden in een bergachtige tropische omgeving. Dat is een uitdaging vanwege het systematisch gebrek aan historische gegevens, de beperkte beschikbaarheid van middelen, en de veelal complexe klimatologische, hydrologische en hydraulische omstandigheden. De bijdragen van dit onderzoek zijn bedoeld om de kennis te vergroten die nodig is voor het ontwikkelen en toepassen van waarschuwingssystemen tegen overstromingsrisico's in stroomgebieden met beperkte hydrologische informatie, rekening houdend met dwarsverbanden tussen waarschuwing en risicobeheersing.

Kennis van de risico's bepaalt het raamwerk voor het ontwikkelen van operationele waarschuwingssystemen tegen overstromingen. In dit onderzoek wordt een methode voorgesteld die op stroomgebied niveau nagaat welke delen gevaar lopen bij snel opkomende overstromingen en waar de ophoping van puin dit proces verergert. De methode is erop gericht om een index te ontwikkelen die verschillende aspecten bestaande uit een morfometrische en een landgebruik component en is toegepast op 106 peri-urbane stroomgebieden in de bergachtige omgeving van Bogotá, Colombia. De index geeft de vatbaarheid van het betreffende (deel)gebied aan als verkregen op basis van informatie die algemeen beschikbaar is in peri-urbane gebieden en kan worden gebruikt om prioriteiten te bepalen voor meer gedetailleerde analyse van bedreigingen in specifieke gebieden. De index geeft aan welk type overstroming verwacht mag worden, hetgeen een cruciale factor is bij het bepalen van de risico's met name in een bergachtige omgeving. Deze kan vervolgens worden gebruikt om de prioriteit te bepalen waar gedetailleerde strategieën moeten worden ontwikkeld teneinde overstromingsrampen te voorkomen. De index is bedoeld om te worden toegepast op regionaal niveau om aan te geven welke deelstroomgebieden mogelijk meer ontvankelijk zijn voor schade ten gevolge van overstromingen dan andere in dezelfde regio.

Voor het vaststellen van het overstromingsrisico is het van belang om zowel de kwetsbaarheid als de gevolgschade te kennen. In dit onderzoek is de gevolgschade bepaald aan de hand van overstromingsanalyses waarna de kwetsbaarheid voor het deelstroomgebied werd verkregen op basis van een Principal Component Analysis van variabelen die volgens de literatuur bijdragen. De kwetsbaarheid index in combinatie met de vatbaarheid index bepaalde vervolgens welke deelgebieden prioriteit kregen. Het complexe samenspel tussen kwetsbaarheid en vatbaarheid kwam duidelijk naar voren bij de casus die in dit proefschrift is onderzocht. Bij toename van milieuproblemen in kwetsbare stroomgebieden blijkt duidelijk dat het overstromingsrisico toeneemt.

Zodra de prioriteitsgebieden zijn vastgesteld voor het ontwikkelen van maatregelen tegen overstromingsrisico's, richt het onderzoek zich op het ontwikkelen van het modelinstrumentarium voor het voorspellen van overstromingen. Een eerste stap betreft het bepalen van de maatgevende neerslag in een complexe topografie die gewoonlijk in bergachtige gebieden wordt aangetroffen. Daarbij zijn verschillende interpolatie-technieken (Ordinary Kriging en Kriging with external Drift) onderzocht die goed in een operationele omgeving van meetstations voor regenval zouden kunnen werken. Ook zijn de prestaties van Pooled Variograms (PV) en de toegevoegde waarde van secundaire variabelen onderzocht. Het onderzoek laat zien dat interpolatietechnieken op basis van PV tot vergelijkbare resultaten leiden als wanneer afzonderlijke neerslag gebeurtenissen worden gebruikt, hetgeen aantoont dat deze benadering geschikt is voor operationele toepassing in het gebied van onderzoek. Ook werden de beperkingen van het gebruik van Kriging with External Drift vastgesteld: deze geeft alleen een verbetering wanneer de aangepaste R2 tussen de secundaire variabelen en de neerslag groter is dan de variabiliteit bij Ordinary Kriging.

De kern van voorspelsystemen voor overstromingen bestaat uit modellen. Daarbij is de keuze van een hydrologisch model van groot belang. Dit is een uitdaging in tropische berggebieden, in het bijzonder in páramos (tropische ecosystemen in berggebieden). Deze worden door de plaatselijke bevolking in Latijns Amerika vaak als heilige gewijde gebieden beschouwd met een belangrijke natuurwaarde. Bovendien fungeren páramos vaak als waterbron voor belangrijke steden in de Andes. Ondanks hun belangrijke rol is de kennis van hydrologische processen in deze gebieden nog steeds erg beperkt. In dit proefschrift is onderzoek gedaan naar het gebruik van een gedistribueerd model (TETIS), een semi-gedistribueerd model (TOPMODEL) en een gelumped model (HEC-HMS inclusief grondvochtigheid) om de afvoer te bepalen van een tropisch páramo regengebied. Door gebruik te maken van de specifieke eigenschappen van de Flow Duration Curves en door de analyse van berekende debieten kon het meest geschikte model worden bepaald voor waarschuwing tegen overstroming. Het effect van rekenroosters is onderzocht voor gedistribueerde en semi-gedistribueerde modellen om te kunnen bepalen welke criteria aan hydrologische voorspelmodellen moeten worden gesteld. De gevoeligheid van de modellen voor variatie in regenval is nagegaan op basis van Gaussische simulaties en de berekende afvoeren werden vergeleken. Daaruit kwam het TOPMODEL als beste naar voren, zij het met een relatief grote spreiding in afvoer.

Numerieke Weersvoorspelling Modellen (NWM) zijn van groot belang om verder vooruit te kijken dan de verzadigingstijd van een bepaald stroomgebied. NWM worden wereldwijd steeds vaker gebruikt door centra voor overstromingsvoorspelling. In dit onderzoek is het WRF model uit Bogotá (Colombia) gebruikt om de toepasbaarheid voor páramo gebieden na te gaan. Daarbij zijn vier strategieën onderzocht om het hydrologisch model in het studiegebied aan te sturen: a) Zero rainfall forecasts; b) Raw WRF forecasts; c) Deterministic bias corrected WRF forecasts; d) Ensemble WRF precipitation forecast. Als referentie voor de afvoervoorspelling is een situatie zonder neerslag gebruikt. De resultaten laten zien dat afvoervoorspelling op basis van een hydrologisch model aangestuurd met nabewerkte WRF neerslaggegevens tot een betere overstromingsvoorspelling leidt vergeleken met de situatie zonder neerslag. Hoewel gebruik van het WRF model slechts tot een bescheiden verbetering leidt, laat deze benadering toch zien dat dit een veelbelovende aanpak lijkt om te komen tot betere voorspellingen in gebieden met hoge meteorologische en topografische complexiteit.

Resumen

Los sistemas de alerta temprana de inundaciones son considerados uno de los instrumentos más efectivos de gestión del riesgo de inundación, cuando están estructurados correctamente dentro de estrategias y políticas integrales de gestión del riesgo. Por consiguiente, se han llevado a cabo muchas iniciativas alrededor del mundo para avanzar en el desarrollo de los componentes que determinan la efectividad de los sistemas de alerta temprana. El objetivo de esta investigación es contribuir al entendimiento de los componentes de conocimiento del riesgo y pronóstico de sistemas de alerta temprana de inundaciones, en el contexto particular de ciudades en desarrollo localizadas en zonas tropicales de alta montaña. Estos componentes implican retos variados teniendo en cuenta la persistente falta de datos, las limitaciones de recursos y generalmente complejas condiciones climáticas, hidrológicas e hidráulicas. Las contribuciones de esta investigación están orientadas al avance en el conocimiento requerido para el diseño y operación de sistemas de alerta temprana de inundaciones en cuencas con escasez de datos, desde una perspectiva hidrológica, sin desconocer la naturaleza transversal de los sistemas de alerta temprana de inundaciones en el proceso de gestión de riesgo de inundación.

El conocimiento del riesgo proporciona el marco base para la operación de sistemas de alerta temprana de inundaciones. En esta investigación, se propone un método regional para evaluación de susceptibilidad a inundaciones y para identificar predisposición a la ocurrencia de flujos de detritos a escala de cuenca. El método se basa en un índice compuesto por un indicador morfométrico y un indicador de cobertura del suelo, el cual es aplicado a 106 cuencas de montaña periurbanas de la ciudad de Bogotá, Colombia. El indicador de susceptibilidad es obtenido de información disponible normalmente encontrada en áreas montañosas periurbanas y puede ser usado para priorizar cuencas que posteriormente pueden someterse a estudios de amenaza más detallados. El indicador es útil para identificar tipos de inundación, que es un paso crucial en la evaluación de riesgo de inundación en zonas montañosas; y puede ser usado como información de entrada para la priorización de estrategias de gestión del riesgo de inundaciones a nivel regional y para la priorización e identificación de análisis de amenaza de inundación detallados. El alcance del indicador es regional y por lo tanto no pretende proporcionar una evaluación detallada, sino identificar las cuencas que podrían potencialmente ser más susceptibles que otras a las inundaciones en la misma región.

El análisis de riesgo requiere la evaluación tanto de la amenaza como de la vulnerabilidad. La amenaza fue obtenida de manera indicativa del análisis de susceptibilidad a las inundaciones y la vulnerabilidad a escala de cuenca fue obtenida posteriormente. La vulnerabilidad fue evaluada con base en un análisis de componentes principales llevado a cabo con variables reconocidas en la literatura como contribuyentes de la vulnerabilidad. Una vez se obtuvo un indicador de vulnerabilidad, este fue combinado con el indicador de susceptibilidad, proporcionando como resultado un índice que permite la priorización de las cuencas como información base para la gestión del riesgo de inundaciones a nivel regional. La compleja interacción entre la vulnerabilidad y la amenaza se evidencia en el caso de estudio. La degradación ambiental en cuencas vulnerables muestra la influencia que la vulnerabilidad ejerce sobre la amenaza y viceversa, estableciendo de esta forma un ciclo de construcción de condiciones de riesgo.

Una vez se identificaron las cuencas prioritarias con base en el análisis de riesgo, la investigación sigue los pasos de modelación para el desarrollo del pronóstico. Como primer paso, la precipitación es abordada en el contexto de la complejidad topográfica comúnmente encontrado en áreas tropicales de montaña. Se analizó la diferencia de funcionamiento de técnicas de interpolación (kriging ordinario y kriging con deriva externa) con el fin de proponer un procedimiento en tiempo real para obtener campos de precipitación utilizando datos puntuales medidos. Se estudió el funcionamiento de variogramas agregados y el valor añadido de variables secundarias en el procedimiento de interpolación. Los resultados mostraron que los interpoladores que usan variogramas agregados proporcionan un funcionamiento comparable a cuando los interpoladores fueron aplicados a tormentas individuales, mostrando que pueden ser usados exitosamente para interpolación en tiempo real en el área de estudio. Adicionalmente, el análisis identificó limitaciones en el uso de kriging con deriva externa. El kriging con deriva externa proporcionó una mejora consistente, solo cuando la correlación entre las variables secundarias y la precipitación es más alta que el porcentaje de variabilidad explicada encontrada en el kriging ordinario.

Los modelos son el corazón de los sistemas de pronóstico de inundaciones. La elección entre los posibles modelos hidrológicos es un aspecto clave, que constituye un reto en áreas montañosas tropicales, particularmente en páramos (ecosistemas tropicales de alta montaña). Los páramos han sido considerados lugares sagrados por la población indígena de Latinoamérica y son reconocidos como áreas de inmenso valor natural. Adicionalmente, los páramos son la fuente de agua de muchas ciudades importantes en los Andes. En contrate con su gran importancia, el conocimiento de sus procesos hidrológicos es aún bastante limitado. En esta investigación, un modelo distribuido (TETIS), un modelo semidistribuido (TOPMODEL) y un modelo agregado (HEC HMS soil moisture accounting) fueron utilizados para simular los caudales de una cuenca tropical de alta montaña con una cuenca alta constituida por zona de páramo. Se llevaron a cabo análisis de funcionamiento y diagnóstico utilizando las señales de la curva de duración de caudales y los flujos de los modelos, con el fin de identificar el modelo más apropiado para el área de estudio con fines de alerta temprana de inundaciones. El impacto del tamaño de celda fue explorado en el modelo distribuido y en el semidistribuido para proporcionar información sobre el criterio a ser usado para modelamiento con fines de pronóstico. La sensibilidad de los modelos a la variación de la precipitación de entrada fue analizada ejecutando los modelos con un ensamble de precipitación obtenido mediante simulación gausiana. Los ensambles de caudal resultantes de cada modelo fueron comprados con el fin de identificar las diferencias entre las estructuras de los modelos. Los resultados muestran que el TOPMODEL es el modelo más realista de los tres que fueron evaluados, mostrando al mismo tiempo la más alta variabilidad en el ensamble de caudal.

Los modelos numéricos de predicción del clima (NWP) son fundamentales para extender el tiempo de anticipación de las alertas más allá del tiempo de concentración de una cuenca. Los NWPs están siendo cada vez más usados en los centros de pronóstico alrededor del mundo. En esta investigación, el modelo WRF bajo la configuración utilizada actualmente por la Agencia Meteorológica Nacional para emitir pronósticos del clima en Bogotá (Colombia), fue usado para explorar su valor agregado para alertas tempranas de inundación en un área de páramo. Se utilizaron pronósticos generados bajo cuatro estrategias para ejecutar el modelo hidrológico del área de estudio: a) Pronósticos de precipitación iguales a cero; b) pronósticos crudos del WRF; c) Pronósticos del WRF con sesgo corregido determinísticamente; y d) pronósticos ensamblados de precipitación obtenidos del WRF. Con el fin de evaluar los pronósticos de caudal obtenidos del modelo hidrológico ejecutado con la precipitación del modelo WRF, se utilizó un pronóstico de referencia equivalente al obtenido del pronóstico de precipitación igual a cero. Los resultados mostraron que los pronósticos de caudal obtenidos del modelo hidrológico ejecutado con la precipitación pos procesada obtenida del modelo WRF tiene un valor agregado para el sistema de alerta temprana cuando se compara con la obtenida de pronósticos de precipitación iguales a cero. A pesar de que el valor agregado de los pronósticos del modelo WRF es modesto, éste es promisorio para incrementar la habilidad del pronóstico en áreas de alta complejidad meteorológica y topográfica y muestra potencial ante la posibilidad de mejoramiento.

Acknowledgements

I would like to thank my copromotor, Dr. Micha Werner, who has been supportive since the day I first proposed a research project on flood early warning in Bogotá. Thank you for your interest and encouragement from the beginning, for your challenging questions, interesting discussions and guidance through the rough road to finish this thesis.

My sincere thanks to Prof. Arthur Mynnet, for his support to successfully finalize this process. Thanks to Prof. Nigel Wright for his valuable input and positive disposition.

Thanks to Prof. Nelson Obregón for his academic and also moral support during all this time. Thanks for your encouragement and kindness. I greatly appreciate the support received from the Geophysiscal Institute of the Javeriana University, directed by Prof. Obregón, for providing the computer resources needed for the hydrological modelling of this research.

A special acknowledgement goes to the Instituto Distrital de Gestión de Riesgo (IDIGER), formerly Fondo de Prevención y Atención de Emergencias (FOPAE), for their support to carry out this research, and to my FOPAE colleagues and friends for the knowledge that they have shared with me.

Thanks to all my friends in the Netherlands that made me feel at home during my busy and short stays. I extend my gratitude to all of them and their families.

I am very much indebted to my family, my husband Juan Miguel Sosa, who supported me in every possible way to see the completion of this work. Thanks to my parents for their love and encouragement in all my pursuits. And thanks to my brother, whose advice and support helped me to persevere in the difficult moments. *Gracias a mi familia por todo su amor y apoyo en estos años. Gracias papá y mamá por estar siempre presentes con sus consejos y cariño. Gracias a mis hermanos por su apoyo en los momentos difíciles y a Laurita, mi sobrina, por su inmensa y contagiosa alegría, que me llevaba a hacer una pausa del trabajo los fines de semana. Gracias Juan por tu paciencia y amoroso apoyo durante estos años.*

Contents

List of Figures

List of Tables

Chapter 1

Introduction

1.1 Background

The United Nations Office for Disaster Risk Reduction UNISDR [2009] defines Early Warning Systems as the set of capacities needed to generate and disseminate timely and meaningful warning information to enable individuals, communities and organizations threatened by a hazard to prepare and to act appropriately and in sufficient time to reduce the possibility of harm or loss. This definition encompasses much more than the scientific and technical tools for forecasting and warning issuing [Maskrey, 1997] and transcends to the political and social context. The concept of people-centred early warning systems [Basher, 2006, ISDR, 2006, Maskrey, 1997, Molinari et al., 2013, NOAA and COMET, 2010, UNISDR, 2009] considers four operational components of effective early warning systems, namely: (i) risk knowledge, (ii) monitoring and warning system, (iii) dissemination and communication and (iv) response capability (see Figure 1.1). These components are closely interconnected and a failure in any one of the four key components leads to the failure of the whole system [ISDR, 2006]. As stated by Maskrey [1997] early warning systems are only as good as their weakest link.

Risk knowledge as a first component, provides the framework for the operation of a flood early warning system. This component should be approached holistically, and includes not only hazard but also exposure and vulnerability factors. Risk knowledge shoud aim at reducing risk not controlling hazard [Molinari et al., 2013]. Knowledge about risk scenarios including vulnerability analysis integrating not only physical vulnerability but also social aspects is crucial for the design, implementation and operation of flood early warning. This constitutes the starting point when designing a flood warning system, providing not only information about the characteristics of the hazard but also the assessment of the locations and numbers of people and

FIGURE 1.1: Flood early warning system components

properties at risk from flooding. Vulnerability studies can also highlight where to target effort in public awareness campaigns, develop flood emergency plans, and plan emergency response [Sene, 2008]. According to Maskrey [1997] early warning must include the development of a risk information sub-system capable of monitoring hazard and vulnerability patterns and of generating risk scenarios for a given area at a specific time.

The second component of a flood early warning system is a monitoring and warning system. At the heart of any early warning system there is a model. Therefore the inherent uncertainties to any meteorological, hydrologic and hydrodynamic model are present, thus warnings are in nature probabilistic [Basher, 2006]. According to Krzysztofowicz [1999] the sources of uncertainty associated with a river forecast can be categorized as operational, input, and hydrologic, involving a cascade of models [Faulkner et al., 2007]. Operational uncertainty refers to erroneous or missing data, human processing errors and unpredictable interventions. Input uncertainty is associated with random inputs to the model; and hydrologic uncertainty includes model, parameter estimation and measurement errors [Krzysztofowicz, 1999]. Minimization of uncertainty can be achieved through data assimilation, for which many techniques exist (Kalman filter methods, ensemble Kalman filter methods, particle filter methods and generalized likelihood uncertainty estimation (GLUE) method) [Faulkner et al., 2007].

Flood forecasting systems are not only required to be robust, adaptive, evolving with experience, timely and sufficiently accurate within pre-determined time horizons but must also provide
a usable quantification of the forecasting uncertainty [Todini et al., 2005]. Many operational
flood early warning systems treat the forecast as deterministic; this is even more common in
developing countries where the development of flood early warning is relatively recent and in
some cases it does not rely on the real time operation of sophisticated forecasting systems.

Besides the technical capacity, uncertainty communication is one of the main challenges in
probabilistic forecasting [Faulkner et al., 2007], with this being crucial for effective flood early
warning, involving practitioners, scientists, decision makers and community.

Even if the monitoring and warning component (see Figure 1.1) is probably the most researched
and recognized, experience has shown that high quality forecasts are insufficient to reduce
impacts and losses [Basher, 2006] and that the human factor in early warning systems is very
important. However, due to the high relevance of the monitoring and warning component,
commonly flood early warning tends to be focused on the generation of forecasts neglecting
vulnerability components that are essential to risk reduction [Basher, 2006].

The third and fouth components correspond to dissemination and communication and response
capability. The former implies all the processes needed so the warnings reach those at risk
[ISDR, 2006], and the latter corresponds to the capacity of the affected communities to take
actions that reduce expected damages [Molinari et al., 2013].

Despite the importance of flood early warning systems in flood risk management, it must be
stressed that flood risk reduction strategies should not rely solely on early warning systems
[Maskrey, 1997]. Flood early warning systems created as the only flood risk management
measure can create a false sense of security, thus being counterproductive, increasing rather
than mitigating flood risk [Molinari et al., 2013]. Flood early warning systems should be
considered a last line of defence and not as the only resource. According to Molinari et al.
[2013] flood early warning systems can be seen as a non-structural short-term measure, whose
aim is the treatment of the so-called unmanaged risk. Once risk has been identified and
quantified, possible mitigation measures are considered to reduce the probability of damage.
These can be structural (e.g. levees, retention basins, and debris retention structures) and non-
structural (e.g. insurance incentives, land use planning, building codes). Flood early warning
is a type of non-structural measure that must be integrated in a broader risk management
framework. Furthermore, due to the cross-cutting nature of flood early warning systems in the
risk management process, flood early warning risk information sub-systems can also provide

the information for land use planning on a permanent basis; communication sub-systems can contribute to risk awareness and education; and disaster preparedness sub-systems can be linked to vulnerability reduction strategies [Maskrey, 1997]. At the same time, flood early warning can be part of a larger framework for multi-hazard early warning embedded in a national and local strategy for risk reduction. Therefore, flood early warning should be integrated with the other measures and policies and should improve in time from lessons learned from operation. Thus, performance assessment techniques are important to ensure effectiveness of flood early warning systems [Sene, 2013]. According to Molinari et al. [2013] the evaluation of performance should be aimed at identifying its capacity to reduce damage.

1.2   Scope of the thesis

Developing cities represent a challenge for flood early warning, taking into account the persistent lack of data, limited resources and often complex climatic, hydrologic and hydraulic conditions. Furthermore, efficient decision support and targeted dissemination of information are important needs; in such a way that warnings derived from these systems can properly be understood to provide real protection to those at risk. The lack of hydro-meteorological information is more noticeable in mountainous areas, where forecasting in fast responding catchments poses high demands to data availability. The general lack of accessibility to hydro-meteorological data can be aggravated further by a lack of agreements to efficiently share data between different institutions. A particularly relevant development is the recognition that uncertainty needs to be considered, and there is a tendency to use ensemble prediction systems, (EPS) in such a way that probabilistic forecasts can be used reliably by decision makers. EPS are now in daily operational use by national weather services around the world including Canada, the United States, Australia and Europe [Demeritt et al., 2007].

In this context, there is a further gap between developing and developed countries, not only regarding availability of data, but also in methodologies applied to process, model and handle uncertainty in forecasts. Furthermore, efforts to address the particular issues present in the issuing of warnings in developing cities are scarce. Such is the case with prioritisation of watersheds to focalize flood early warning efforts; appropriate description of the spatial distribution of rainfall in areas with complex topography and meteorology; assessment of hydrological models in tropical high mountain basins; and the potential use of numerical weather models for flood forecasting in tropical high mountain basins.

This research is aimed at contributing to the closing of these gaps, taking into account the particular conditions in developing cities. The starting point is the research of methods to establish a hazard and risk framework that provides the basis for effective implementation of flood early warning. Subsequently, the flood forecasting component is addressed. Dissemination and communication and response capability are beyond the scope of the research. The objective of the research is to develop and demonstrate methods for reliable operational forecasting of flood hazards in developing cities. The central question posed in the research is: *How can a reliable operational flood forecasting system be established in developing cities, considering uncertainty as an effective tool for decision making?*

The following are important aspects in the focus of the research in order to effectively contribute to a better understanding of multi-scale flood early warning in developing cities:

• In developing countries, records of past events are scarce, and the identification and validation of flood hazard areas and risk areas becomes a challenge, particularly when taking into account the dynamic nature of hazard and risk and the fact that these occur at variable time and spatial scales. This research addresses the issue of prioritising a large area with mixed uses (urban and rural) for flood risk management purposes, providing guidance for decision making on areas where measures such as flood early warning should be implemented. The research questions posed to address this aspect are: *When little or no historical information is available, how can hazards produced by debris flows, and by clearwater flows be distinguished using geomorphic data? What physical parameters of the watersheds can be used as reliable indicators of the type of flash flood expected, taking into account highly modified watersheds? Can a robust method to determine hazard areas be developed when several geomorphical characteristics of a flashy basin are not known, and to which extent can the methods be simplified to allow reliable identification of the hazard areas even with little data? Can a prioritisation method be developed in areas with little data, so critical watersheds from a flood risk perspective can be identified?*

• Flash floods are common in both developing and developed cities, and owing to their characteristic space and time scales, there are specific problems to monitor and predict these. These events generally develop at space and time scales that conventional measurement networks of rain and river discharges are not able to sample effectively. Flash flood monitoring requires rainfall estimates at small spatial scales (1 km or finer) and short time scales (15-30 minutes, and even less in urban areas). These requirements are generally met by weather radar networks, but these instruments are not common in developing cities. This research explores methodologies to produce rainfall fields in real time based on rainfall gauges and the associated uncertainty. The research question posed to address this aspect is: *What secondary variables apart from precipitation can be incorporated into the rainfall model to improve the interpolation of precipitation at different time scales?.*

• Developing cities located in the Andes, face special hydrological issues particular to high mountain environments, referred to in the Andes as Páramo. Páramos constitute special ecological and hydrological zones that can be found in northern Peru, Ecuador, Colombia and Venezuela. The water quality in these areas is excellent, and the rivers descending from the páramo provide a high and sustained base flow, which is an important source of water for many developing cities in the countries mentioned. An important characteristic of these areas is the extremely high water retention [Célleri and Feyen, 2009]. Very few attempts to model the páramo hydrology are known. This is no surprise, given the scarcity of long term hydrological data sets. However, also from a conceptual viewpoint, the description and modelling of the hydrological processes in these soils is particularly challenging. In this research, a páramo area was chosen to test modelling approaches for flood early warning. The area corresponds to the upper basin of one of the main rivers in Bogotá Colombia. The research question posed to address this aspect is: *What is the most appropriate modelling approach for a páramo watershed?, In the case of distributed and semi-distributed models what grid size should be used for appropriate representation of hydrological processess?* .

• The advance in scientific understanding of not only the physical processes that allow flood threat to be anticipated on, but also on uncertainties and how best to deal with these to provide optimal decision support constitute an important need in developing cities. In this research input and model uncertainty are explored in the context of mountain watersheds. The research question posed to address this aspect is: *What is the importance of input and model uncertainty in the modelling results of a páramo watershed?.*

• Numerical Weather Prediction (NWP) models are fundamental to extend lead-times beyond the concentration time of a watershed. However, their results contain noise, are contaminated by model biases, are too coarse to adequately resolve all features such as convection, and are influenced by uncertainty inherent in the initial conditions [Colman et al., 2013]. Furthermore, weather forecasting in tropical mountains is highly challenging due to meteorological complexity and lack of monitoring data. In this research the potential of an NWP model is assessed for flood forecasting in a páramo área. The research question posed to address this aspect is: *What is the added value of an NWP model in a flood forecasting system in a páramo area? and what possible improvement can bias correction procedures provide?.*

The case study for this research is Bogotá Colombia, where a dense hydrologic network, a high topographical variability and complex climatic conditions under the typical conditions of a developing city take place. This city provides a perfect scenario for analysis and development of methods.

1.3   Outline of the thesis

This thesis is drawn from five papers published or under review. Chapters 2 and 3 present a methodology for regional prioritisation of flood risk in mountainous watersheds. Chapter 2 presents a method for assessing regional debris flow susceptibility at the watershed scale, based on an index composed of a morphometric indicator and a land cover indicator. The indicator of debris flow susceptibility is useful in the identification of flood type, which is a crucial step in flood risk assessment especially in mountainous environments; and it can be used as input for prioritisation of flood risk management strategies at regional level and for the prioritisation and identification of detailed flood hazard analysis.

Chapter 3 focuses in the regional analysis of flood risk carried out in the mountainous area surrounding the city of Bogotá. Vulnerability at regional level was assessed on the basis of a principal component analysis carried out with variables recognised in literature to contribute to vulnerability; using watersheds as the unit of analysis. The complex interaction between vulnerability and hazard is evidenced in the case study. Environmental degradation in vulnerable watersheds shows the influence that vulnerability exerts on hazard and vice versa, thus establishing a cycle that builds up risk conditions.

Chapter 4 presents the research carried out to propose a method to produce rainfall fields in real time for flood early warning purposes. The differences in performance of Ordinary Kriging, Universal Kriging and Kriging with External Drift with individual and pooled variograms were assessed for 139 daily datasets with significant precipitation in the study area. The analysis identified limitations in the use of Kriging with External Drift and the differences between interpolation methods and their significance.

Chapter 5 explores the performance of a distributed model (TETIS), a semi-distributed model (TOPMODEL) and a lumped model (HEC HMS soil moisture accounting) in the upper area of the basin that contains most of the prioritary watersheds identified in chapter 3. The impact of varying grid sizes was assessed in the TETIS model and the TOPMODEL, in order to chose a model with balanced model performance and computational efficiency. Differences of performance among model structures are studied in comparison with the uncertainty of the precipitation input.

Chapter 6 takes as starting point the TOPMODEL described in chapter 5 to study the added value of the precipitation forecasts produced with the Weather Research and Forecasting Model (WRF). Different post processing strategies to produce forecasts from the WRF model are studied through the use of verification techniques.

Chapter 7 presents the conclusions and recommendations.

Chapter 2

Regional debris flow susceptibility analysis in mountainous peri-urban areas through morphometric and land cover indicators

This chapter is an edited version of: Rogelis, M. C. and Werner, M.: Regional debris flow susceptibility analysis in mountainous peri-urban areas through morphometric and land cover indicators, Nat. Hazards Earth Syst. Sci., 14, 3043-3064, doi:10.5194/nhess-14-3043-2014, 2014.

2.1   Introduction

Appropriate recognition of hydrogeomorphic hazards in mountain areas is crucial for risk management, since it provides the basis for more detailed studies and for the development of appropriate risk management strategies [Jakob and Weatherly, 2005, Welsh, 2007, Wilford et al., 2004]. Besides the identification of the flood potential, it is important to distinguish between debris-flow and non debris-flow dominated watersheds since these constitute very different hazards.

There are several definitions for hydro-geomorphic processes. Wilford et al. [2004] distinguishes among floods, debris floods and debris flows with sediment concentrations of 20% and 47% as upper limits for floods and debris floods respectively. Santangelo et al. [2012] and Costa [1988] differentiate water floods as newtonian, turbulent fluids with non-uniform concentration profiles and sediment concentrations of less than about 20% by volume and shear strengths less than 10 N/m^2; hyperconcentrated flows as having sediment concentrations ranging from 20 to 47% by volume and shear strengths lower than about 40 N/m^2; and debris flows as being non-Newtonian visco-plastic or dilatant fluids with laminar flow and uniform concentration profiles, with sediment concentrations ranging from 47 to 77% by volume and shear strengths greater than about 40 N/m^2. On the other hand, FLO-2D Software [2006] uses the terms mudflow (non-homogeneous, non-Newtonian, transient flood events), and mud flood (sediment concentration from 20% to 40-45% by volume). Despite the variety of definitions, the characteristics of debris flows imply different hazard conditions from those related to clear water floods, with debris flows being potentially more destructive. The higher destructive capacity is related to a much faster flow and higher peak discharges than those of a conventional flood; as well as high erosive capacity with the ability to transport large boulders and debris in suspension and the generation of impact forces comparable to rock and snow avalanches [Santangelo et al., 2012, Welsh, 2007]. With a lower sediment concentration, debris floods and hyperconcentrated flows as presented by Wilford et al. [2004] and Santangelo et al. [2012] are less hazardous, since they carry less of the large boulders responsible for impact damage in debris flows and flow velocities are usually lower. They are, however, considered more dangerous than clear water floods of similar magnitude [Welsh, 2007]. Previous research on the identification of flood potential and areas susceptible to debris flows used quantitative methodologies such as logistic regression and discriminant analysis in addition to GIS and remote sensing technologies [Bertrand et al., 2013, Chen and Yu, 2011, Crosta and Frattini, 2004, De Scally et al., 2010, De Scally and Owens, 2004, Griffiths et al., 2004, Kostaschuk, 1986, Patton and Baker, 1976, Rowbotham et al., 2005, Santangelo et al., 2012, Wilford et al., 2004]. These studies focused on the identification of basins or fan parameters to classify them according to their dominant hydro-geomorphic processes. A conclusion from these studies is that drainage basin morphology is an important control of fan processes [Crosta and Frattini, 2004] and that there are significant differences in morphometric characteristics between basins where the dominant process is debris flows and those mainly dominated by fluvial processes [Welsh, 2007]. Morphometric parameters such as the basin area, Melton ratio and watershed length have been identified by several authors as reliable predictors for differentiating between debris-flow and non-debris-flow dominated watersheds and their respective fans [Welsh, 2007]. However, the results of the analyses seem to be highly dependent on the geographical area where the methodology is applied and in many cases the identification of morphometric parameters requires a previous independent classification of the watersheds normally entailing stratigraphic observations, detailed field work, aerial photo analyses and calculations.

When historical data on the occurrence of flash floods and debris flows are not available, the recognition of hydro-geomorphological hazards can be carried out through field work analysis applying methods such as the one proposed by Aulitzky [1982] based on hazard indicators, or through stratigraphic evidence in conjunction with age control [Giraud, 2005, Jakob and Weatherly, 2005]. However, such fieldwork and detailed geological and geotechnical analysis at the regional scale require significant resources and time, and may not be practicable in the extensive peri-urban areas of cities in mountainous areas such as those in the Andean cordillera. Furthermore, urbanisation processes in the peri-urban areas of these cities make geologic investigation difficult. Moreover the history of the watershed may not be a conclusive indicator of current hazard conditions, since anthropogenic intervention can play a significant role in the hazard dynamics. This calls for a more rapid yet reliable assessment of the watersheds, allowing a prioritization of watersheds where a more detailed analysis based on field data is to be carried out.

This study proposes a method for regional assessment of debris flow susceptibility under limited availability of data in urban environments, where flash floods occur as debris flows, hyperconcentrated flows or clear water flows as defined by Costa [1988]. The proposed index is based exclusively on information derived from digital elevation models and satellite images to overcome the limitation often found in the availability of previous geological work such as stratigraphic analysis and fieldwork for large areas.

The ability of morphometric variables to identify debris flow dominated basins was tested. Morphometric variables and land cover characteristics were considered as factors that influence flood hazard, and were combined in an index that can be interpreted as the potential susceptibility to which watersheds are prone, including the spatial differentiation of the dominant type of hazard. A key aspect of the index is the discrimination between debris flow and clear water flood dominated watersheds in order to understand the level of threat that floods in the watersheds pose, and to support prioritisation of watersheds to be subjected to further detailed study.

The study area is the mountainous area surrounding the City of Bogotá (Colombia), where an accelerated urban process has taken place during the last decades, forming a peri-urban area mostly characterised by illegal developments. To overcome the lack of historic records and the infeasibility to carry out detailed geologic fieldwork for the identification of hydrogeomorphological processes that allow validation of the susceptibility index, results are compared with an independent method based on the propagation of debris flows using a digital elevation model as well as with the few available flood records in the area.

2.2 Methods and Data

2.2.1 Study Area

This research focuses on the mountainous watersheds surrounding the city of Bogotá, the capital and economic centre as well as the largest urban agglomeration of Colombia with an estimated 7.4 million inhabitants. The city is located in the Andean region (see Figure 2.1). Several creeks drain the steep mountains surrounding the city and cross the urban area to finally drain into the larger Bogotá River. In this analysis the watersheds that drain into the main stream of the Tunjuelo river basin, one of the largest tributaries of the Bogotá River, as well as the watersheds in the Eastern Hills were considered.

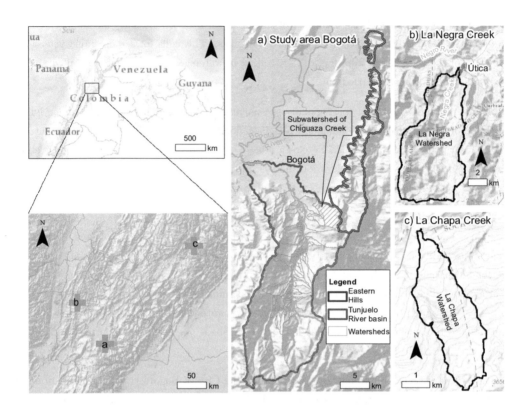

FIGURE 2.1: Location of the study areas. Service Layer Credits: Sources: Esri, HERE, DeLorme, Intermap, increment P Corp., GEBCO, USGS, FAO, NPS, NRCAN, GeoBase, IGN, Kadaster NL, Ordnance Survey, Esri Japan, METI, Esri China (Hong Kong), swisstopo, MapmyIndia, OpenStreetMap contributors, and the GIS User Community

This includes 66 watersheds in the Tunjuelo River basin and 40 in the Eastern Hills of Bogotá (see Figure 2.1-a). These are characterised by a mountainous environment with areas ranging between 0.2 km^2 and 57 km^2. The area is mainly formed by sandstones of Cretaceous and Palaeogene age. This sedimentary rock forms surrounding mountains up to 4000 m altitude, thus reaching to some 1500 m above the level of the high plain of Bogotá [Torres et al., 2005]. The mean annual precipitation varies from 600 mm to 1200 mm in a bimodal regime with rainy seasons in April-May and October-November [Bernal et al., 2007].

In the study area, flooding is controlled by climate and physiography. However, land use practices and anthropogenic influences have increased flood risk not only through soil and land cover deterioration but also through an intensive occupation of floodplains. In the southern mountains of Bogotá, which belong to the Tunjuelo river basin, the urban and industrial growth has been accelerated from the 1950s. Between 1951 and 1982, the lower basin of the Tunjuelo river was the most important area for urban development. It was settled by the poorest population of Bogotá [Osorio, 2007], and its growth has been characterised by informality and lack of planning. The most devastating floods in Bogotá have occurred in the lower Tunjuelo river basin, involving not only the main stream but also the tributary creeks where flash floods have caused human losses [DPAE, 2003a,b,c]. The watersheds located in the Eastern Hills of Bogotá have different characteristics since most of the area corresponds to protected forests. However, informal urbanisation takes place in some areas. Additionally, mining has been common both in the Tunjuelo river basin and in the Eastern Hills, causing the deterioration of the environmental conditions of the watersheds.

The reaches of the creeks in the urban areas have been subjected to significant intervention and occupation. Most of the creeks in the Eastern Hills drain into the storm water system through structures with low hydraulic capacity (less than the return period of 10 years) (Hidrotec, 1999). The streams that drain into the Tunjuelo river have been severely modified, mainly in the reaches near the confluence, albeit without a comprehensive flood management plan. The flood control structures in the study area have been constructed in the main stream of the Tunjuelo river, including a dry dam in the middle basin, three retention basins in the lower basin and levees. Additionally, there are two reservoirs in the upper basin of the Tunjuelo River that supply water to Bogotá. Conversely, flood control works have not been constructed in any of the watersheds in the study area except for the Chiguaza watershed, where levees and channelization works were constructed in the confluence with the Tunjuelo river in 2008.

For the purposes of this study the area formed by the Tunjuelo river basin and the Eastern hills will be considered as comparable, due to the lithologic composition and homogeneous flood management policy in the city.

In order to test the performance of the proposed morphometric indicator, additional to the watersheds in the study area, a sub-watershed of the Chiguaza Creek in the Tunjuelo river basin (see Figure 2.1-a) was analysed, as well as two areas external to the study area: La Negra Creek and La Chapa Creek (see Figure 2.1-b and Figure 2.1-c). These additional watersheds were included in the analysis given the availability of records and previous studies.

La Negra creek is located 67 km to the North West of Bogotá. Among several records of inundation events, the most critical occurred on the 17th of November 1988. The flow moved along La Negra Creek to its confluence with the Negro River affecting an important area of the municipality of Utica [UNAL and INGEOMINAS, 2007]. According to UNAL and INGEOMINAS [2007] the characteristics of this event reveal a concentration of sediment of 40% by volume, which corresponds to the upper limit of the mud flood category according to FLO-2D Software [2006] or to a debris flood according to Wilford et al. [2004].

La Chapa creek was chosen due to the high frequency of debris flows. Chaparro [2005] notes that La Chapa creek is prone to debris flows characterised by the mobilization of granular material of varying size, ranging from boulders of several meters in diameter to sand, embedded in a liquid phase, formed by water, fine soils and air, sometimes accompanied by vegetal material. The most recent event that took place in this watershed was recorded on video, which allowed the type of flow that dominates the watershed to be confirmed.

2.2.2 Methodology

Variability in the level of hazard is reflected in the proposed susceptibility index, where high values represent a higher potential for debris flow and therefore an increased hazard condition. Moreover, flashier conditions, which result from unfavourable morphometric and land cover conditions, contribute to high values of the index, providing an indication of potential for flash flood danger in a large area. The proposed index to represent the level of debris flow susceptibility at regional scale is composed of a morphometric indicator and a land cover indicator.

The units of analysis correspond to the watersheds delineated up to the confluence with the Tunjuelo River in the case of the streams located in the Tunjuelo river basin, and to the confluence with the storm water system in the case of the streams located in the Eastern Hills.

In order to develop the susceptibility index and identify if it is appropriate, a methodology that can be divided into three stages was followed. The first stage addresses the development of the morphometric indicator, the second stage corresponds to the development of the land cover indicator and the third stage is the development of the susceptibility index. Figure 2.2 shows the main steps that were carried out to obtain the susceptibility index for the study area.

For stage 1, a model to calculate a morphometric indicator was developed by using Principal Component Analysis on morphometric parameters that have been identified in literature as important descriptors of flood potential and debris flow discriminators. Due to the poor availability of historical records in the study area, which can limit the validation of the proposed indicator, three independent methods where used to assess the appropriateness of the morphometric indicator (methods i, ii and iii in Figure 2.2). The first method identifies debris flow source areas using two criteria (a and b in Figure 2.2) and propagates the flow on a digital elevation model (DEM) using two angles of reach (ratio between the elevation difference and length from the debris flow initiation point to the downstream extent of the debris flow runout) [Horton and Jaboyedoff, 2008, Kappes et al., 2011], in order to identify the capacity of watersheds to transport potential debris flows to their fans. The binary result of the propagation reaching or not reaching the fan was used to classify the watersheds. The distribution of the values of the morphometric indicator and its component indicators was analysed grouping the values according to the classification obtained from the propagation results. Furthermore, a contingency table and its proportion correct (fraction of watersheds that were correctly identified by the morphometric indicator) were used to establish the correspondence between the morphometric

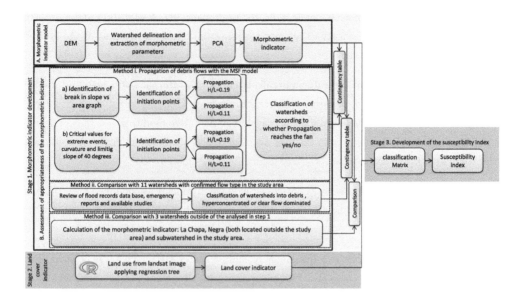

FIGURE 2.2: Schematic representation of the Methodology

indicator and the classification from the propagation modelling to assess the skill of the morphometric indicator to identify the potential capacity of the watersheds to propagate debris flows.

In order to compare the morphometric indicator with field data, method ii was used (see Figure 2.2). A flow type classification of 11 watersheds was carried out on the basis of the available studies, reports and the flood records database managed by the municipality. The flood records database contains 55 flood events from 2001 to 2012. Due to the short period of record of the database, robust frequency analysis is not feasible. Moreover, flood records are less frequent in the Eastern Hills and non-existent in the upper Tunjuelo river basin, which may be due to the low density of population in this latter area. However, the data contained in the database normally describes affected people, type of flow and damage, and provides relevant recent historical information on the type of hydrogeomorphic processes that take place in the watersheds. Watersheds where reports, studies or flood records clearly identify the occurrence or imminent possibility of debris flows were classified as debris flow watersheds (D), watersheds where a significant sediment concentration was identified in the past floods were classified as hyperconcentrated flow watersheds (H) and watersheds where the available reports describe the occurrence of floods without description of sediment sources and sediment concentration were classified as clear water flow watersheds (C).

The correspondence between the morphometric indicator and the classification obtained from flood records, studies and reports in the study area was assessed through a contingency table.

The two contingency tables (morphometric indicator vs propagation classification and morphometric indicator vs flood records classification) allowed assessing the representativeness of the indicator in terms of debris flow threat level.

Additionally, the morphometric indicator was calculated for two external watersheds and a subwatershed of Chiguaza creek in the study area. This constitutes method iii in Figure 2.2. Since information of the dominant processes of these watersheds is available, they were used to assess the applicability of the indicator outside the study area in the first case and to add a valuable information to the analysis of the study area in the second case.

A qualitative indicator of land cover was developed in stage 2, which was combined with the morphometric indicator through a classification matrix and assessed through contingency tables in stage 3 (see Figure 2.2) .

The main input for the methods is a five-meter resolution raster DEM. This was constructed using contours that in the peri-urban area are available at intervals of 1 meter. The contours were processed to obtain a triangulated irregular network that was subsequently transformed into a raster through linear interpolation. The details of each stage of the process are described in the following subsections.

2.2.2.1 Development of the morphometric indicator

Morphometric parameters used in literature (see Table 2.1) were extracted for each watershed from the digital elevation model of the study area using ArcGis, SAGA and R functions. Many of the variables as listed in Table 2.1 are closely correlated. To reduce the dimensionality of the data set, principal component analysis was applied. A reduction of the variables is achieved by transforming the original variables to a new set of variables, the principal components, which are uncorrelated and which are ordered according to the components that retain most of the variation present in the original set of variables (Jolliffe, 2002). These transformed variables were subsequently used to obtain the morphometric indicator.

Differentiation of debris flow watersheds or fans from those dominated by clearwater floods has been carried out by several authors, finding that morphometric variables are very valuable as discriminators of processes in watersheds [Bertrand et al., 2013, Chen and Yu, 2011, Crosta and Frattini, 2004, De Scally et al., 2010, De Scally and Owens, 2004, Griffiths et al., 2004, Jackson et al., 1987, Kostaschuk, 1986, Patton and Baker, 1976, Rowbotham et al., 2005, Santangelo et al., 2012, Wilford et al., 2004]. On the other hand, research on the relationships between watershed characteristics and peak-flood and flashiness has contributed to identify morphometric variables that can help to describe the characteristics of the hydrologic response of a watershed [Patton, 1988]. Table 2.1 summarizes the morphometric variables that have been identified in literature as appropriate discriminators of processes and descriptors of the hydrologic response of watersheds and that were chosen for the analysis.

TABLE 2.1: Morphometric variables used in the analysis. Note that L corresponds to the length of the streams in a watershed, H_{max} and H_{min} correspond to the highest and lowest elevation in a watershed respectively

Variable	Relevance	Reference
Area (A)	Correlated with discharge; proportional to sediment storage in the catchment; wide basins collect a large amount of water, which can dilute the flood reducing the probability of debris flow. Correlated with other morphometric parameters.	Crosta and Frattini [2004], Baker [1976], De Scally and Owens [2004], Gray [1961], Shreve [1974]
Perimeter (P)	Base for several watershed shape indices.	Zavoianu [1985]
Drainage density $DrD = \sum_{i=1}^{n} L_i / A$	Correlated with base flow, peak flood discharge and flood potential.	Baker [1976], Patton and Baker [1976]
Watershed length (L_{wshd})	Has been used to differentiate between watersheds prone to debris flows and debris floods in combination with the Melton ratio	Wilford et al. [2004]
Watershed mean slope (S)	Related to flashiness of the watershed. Used to discriminate between debris flow and clearwater flood domiunated watersheds.	Al-Rawas and Valeo [2010], de Matauco and Ibisate [2004]
Main stream slope (StrS)	Used to discriminate between processes in watersheds.	Welsh [2007]
Relief ratio $R_{Ra} = (H_{max} - H_{min}) / L_{wshd}$	Used to describe debris flow travel distance and event magnitude.	Chen and Yu [2011]
Shape factor $SF = A / L_{wshd}$	Related to flow peak and debris flow occurrence	Chen and Yu [2011], Al-Rawas and Valeo [2010], Wan et al. [2008]
Main stream length (L_{str})	Used to discriminate between processes in watersheds.	Chen et al. [2010]
Circularity coefficient $C = 4\pi * A / P^2$	The more circular a watershed is, the sharper its hydrograph, increasing flashiness and therefore the threat of flooding.	de Matauco and Ibisate [2004]
Elongation ratio $E = 2/(L_{wshd}(A/\pi)^{0.5})$	Floods travel less rapidly; have less erosion and transport potential; and less suspended load in elongated watersheds.	Zavoianu [1985]
Watershed width $W_{wshd} = A / L_{wshd}$	Related to the size of fans	Weissmann et al. [2005]
Length to width ratio (LW)	Measure of elongation	Zavoianu [1985]
Melton ratio $M = (H_{max} - H_{min}) / A^{0.5}$	Frequently used to discriminate among hydrogeomorphologic processes.	Welsh and Davies [2010], Sodnik and Miko [2006], Saczuk [1998], Rowbotham et al. [2005], Wilford et al. [2004]
Hypsometric integral (HI)	Linked to the stage of geomorphic development of the basin; indicator of the erosional stage; related to several geometric and hydrological properties such as flood plain area and potential surface storage; empirical correlations have been established between the hypsometric parameters and observed time to peak. Used to differentiate between processes in the watershed.	Pérez-Peña et al. [2009], Harlin [1978], Luo and Harlin [2003], Willgoose and Hancock [1998], Hurtrez et al. [1999]
Hypsometric skewness (Hs)	Reflects the amount of headward erosion attained by streams; high values are characteristic of headward development of the main stream and its tributaries, representing the amount of headward erosion in the upper reach of a basin.	Harlin [1984]
Hypsometric kurtosis (Hk)	Large values signify erosion in both upper and lower reaches of a basin.	Harlin [1978]
Density skewness (DHs)	Indicates where slope changes are concentrated, and if accelerated forms of erosion, like mass wasting, are more probable in the basin's upper reaches. When density skewness equals 0 equal amount of change is occurring, or has occurred, in the upper and lower reaches of the watershed.	Harlin [1984]
Density kurtosis (DHk)	Relates to the mid-basin slope.	Harlin [1984]
Average of the multiresolution index - MRI Mean (MRIm)	Discriminates between depositional regions and erosional regions.	Gallant and Dowling [2003]

The parameters that are most commonly found in literature as important discriminators of hydrogeomorphic processes are the area, the slope and the Melton ratio. However, other parameters such as those derived from the hypsometric curve and the average of the multiresolution index [Gallant and Dowling, 2003] have also been included given their importance in the description of the evolution and erosion processes of watersheds in the case of the former and the description of the erosion areas in the case of the latter.

The hypsometric curve and the hypsometric integral are non-dimensional measures of the proportion of the catchment above a given elevation [Willgoose and Hancock, 1998]. The hypsometric curve describes the landmass distribution and thus the potential energy distribution within the basin above its base [Luo and Harlin, 2003]. This curve can be seen as an exceedence distribution of normalised elevation where the probability of exceedence is determined by the portion of the basin area that lies above the specified elevation [Huang and Niemann, 2008]. The hypsometric integral is defined as the area below the hypsometric curve. Values near to 1 in the hypsometric integral indicate a state of youth and are typical of convex curves. Nevertheless, mature s-shaped hypsometric curves can present a great variety of shapes, but have the same hypsometric integral value [Pérez-Peña et al., 2009]. In order to analyse the hypsometric properties of the watersheds, the procedure described by Harlin [1978] was used: the hypsometric curve was treated as a cumulative distribution function. The second, third and fourth moments were derived about the centroids, yielding measures of skewness and kurtosis for the hypsometric curves, which are represented by a continuous third order polynomial function.

The multiresolution valley bottom flatness index (MRI) is obtained through a classification algorithm applied at multiple scales by progressive generalisation of the DEM combined with progressive reduction of the slope class threshold. The results at different scales are then combined into a single index. The MRI utilizes the flatness and lowness characteristics of valley bottoms. Flatness is measured by the inverse of the slope, and lowness is measured by a ranking of the elevation with respect to the surrounding area. The two measures, both scaled to the range 0 to 1, are combined by multiplication and can be interpreted as membership functions of fuzzy sets [Gallant and Dowling, 2003].

From the principal component analysis of the morphometric variables, the factor loadings, which represent the proportion of the total unit variance of the indicator which is explained by the principal component, were used to construct the weights of the indicators [Nardo et al., 2008]. In order to develop an overall morphometric indicator the individual indicators obtained from the principal components were combined using as weights the variability explained by each principal component.

The appropriateness of the morphometric indicator to capture the level of debris flow susceptibility was assessed through its comparison with the debris flow propagation capacity of the watersheds; with the classification of 11 watersheds from available detailed studies and historic information; and through the analysis of the indicator obtained in two watersheds outside the study area and one subwatershed of the study area where debris flows have been confirmed. For the first two analyses contingency tables were used and for the third direct comparison of the values of the indicator was carried out.

Several hypotheses have been formulated to explain mobilisation of debris flows and this aspect represents an active research field. The triggering mechanisms and the causal relationships are, however, still partially unknown [Salvetti et al., 2008]. Approaches for the identification of debris source areas include the use of credal networks [Antonucci et al., 2007], the use of indices for predisposition factors to assess debris-flow initiation hazard [Bonnet-Staub, 2000], empirical relationships [Baumann and Wick, 2011, Blahut et al., 2010, Horton and Jaboyedoff, 2008], the Melton's Ruggedness Number [Rengifo, 2012] and the use of the slope versus area diagram as a topographic signature of debris flow dominated channels [Santos and Duarte, 2006]. Two of these approaches to identify potential debris flow initiation points will be used in this study for method i in Figure 2.2. The first approach is based on the analysis of the break in the slope versus drainage area relationship, while the second uses an empirically determined critical condition in this relationship [Horton and Jaboyedoff, 2008]. In both cases, the debris flow propagation areas were obtained through a propagation algorithm by considering two angles of reach (ratio between the elevation difference H and length from the debris flow initiation point to the downstream extent of the debris flow runout L) [Horton and Jaboyedoff, 2008, Kappes et al., 2011].

Regarding the first method of identification of debris source areas, the slope–area diagram is the relationship between the slope at a point versus the area draining through that point. It quantifies the local topographic gradient as a function of drainage area. Several authors have found a change in the power-law relationship (or a scaling break) in slope–area data from DEMs at the point that the valley slope ceases to change below a certain drainage area. This has been inferred to represent a transition to hillslope processes and has been interpreted as the topographic signature for debris flow valley incision [Montgomery and Foufoula-Georgiou, 1993, Seidl and Dietrich, 1993, Stock and Dietrich, 2003]. The same conclusion was made by Tucker and Bras [1998] explaining that different processes have an impact on the slope–area relationship, suggesting the possibility that slope–area data may be used to discriminate between different geomorphic process regimes.

Two distinct regions of the slope–area diagram are typically observed. Small catchment areas are dominated by rainsplash, interrill erosion, soil creep or other erosive processes that tend to round or smooth the landscape. As the catchment area becomes larger, a break in gradient of the curve occurs. This is where slope decreases as catchment area increases. This region of the catchment is dominated by fluvial erosive processes that tend to incise the landscape [Hancock, 2005].

The slope–area curve was constructed for two regions of the study area corresponding to the Tunjuelo river basin and to the Eastern Hills of Bogotá. The break in the slope–area diagram was obtained using segmented regression, in order to determine a threshold to differentiate two regions, one dominated by erosive processes and the other dominated by fluvial erosive processes. This threshold will be used as the topographic signature of debris flow.

In the case of the second method that was applied to identify debris flow sources, corresponding to the procedure proposed by Horton and Jaboyedoff [2008], this applies criteria based on area, slope, curvature, hydrology, lithology and land cover. The slope criterion to identify debris flow source areas, is based on the relationship between slope and drainage area shown in Equation 2.1 and Equation 2.2, where β_{lim} is the threshold slope in degrees and S_{UA} is the upstream area in km^2. These equations were built on observations made by Rickenmann and Zimmermann [1993]. Horton and Jaboyedoff [2008] denominated these criteria as threshold for extreme events given that the 1987 events on which the threshold is based, were considered as extraordinary and this denomination allowed differentiation from other set of points used by Horton and Jaboyedoff [2008].

$$Tan\beta_{lim} = 0.31S_{UA}^{-0.15} \quad if \ S_{UA} < 2.5km^2 \tag{2.1}$$

$$Tan\beta_{lim} = 0.26 \quad if \ S_{UA} >= 2.5km^2 \tag{2.2}$$

In the method by Horton and Jaboyedoff [2008] every point located above the limits defined by equation 2.1 and equation 2.2 is considered as critical. In the application of the method in Argentina [Baumann and Wick, 2011], the equations were bounded between 15 degrees and 40 degrees since in the observations made by Rickenmann and Zimmermann [1993] in Switzerland all the triggering areas slope angles were below 40 degrees and contributing areas inferior to 1 ha were not considered as potential sources. Thus, the parameters used for detection of triggering areas are slopes in the range of 15-40 degrees, contributing areas superior to 1 ha and plane curvatures inferior to -0.01/200 m^{-1} under the condition that the point is located above the limit defined by Equation 2.1 and Equation 2.2.

Using the threshold obtained from the analysis of the slope–area curve together with the criteria of curvature and minimum drainage area as proposed by Horton and Jaboyedoff [2008] initiation points were identified in the study area constituting the first method of identification of debris flow initiation points. As a second method, the threshold of extreme events was used as a criterion for slope and area and in addition the minimum area and curvature were used as recommended by Horton and Jaboyedoff [2008]. The Modified Single Flow Direction (MSF) model [Gruber et al., 2009] was used to identify the areas that potentially could be affected by debris flows for the two groups of initiation points. The MSF is based on the single flow direction (D8) algorithm and other standard functionalities of ArcInfo/ArcGIS to account for flow spreading allowing the flow to divert from the steepest descent path by as much as 45 degrees on both sides. The only required inputs are the source areas and a DEM. For a detailed explanation on the MSF algorithm see Gruber et al. [2009]. As a stopping condition the MSF algorithm uses the angle of reach. The trajectory component of the MSF model usually provides the potential maximum inundation zones of a mass-movement event. Thus, it indicates which areas are more or less likely to be affected. However, the runout distance should also be based on a maximum. A reasonable angle of reach (H/L ratio) has to be evaluated on the basis of empirical data for the type of mass movement that is being modelled. Several efforts have been made to develop relationships to estimate the angle of reach mainly using the volume of the debris flow. The minimum angle of reach that has been observed is 6.5 degrees (ratio H/L=0.11) [Prochaska et al., 2008] and the highest and more repetitive is 11 degrees (ratio H/L=0.19) [Huggel et al., 2003, Kappes et al., 2011, Rickenmann, 1999, Rickenmann and Zimmermann, 1993]. The two angles were used to test the sensitivity of the results but, larger and more fluid debris flows may show lower H/L ratios and consequently a larger flow reach.

Watersheds where the propagation area reaches the mouth of the drainage area using a ratio H/L of 0.19, are classified as debris flow dominated and labelled "0.19H/L". Watersheds where the propagation reaches the mouth for a ratio H/L of 0.11, will be considered debris flow dominated as well, albeit with a more fluid flow. These are labelled "0.11H/L". In this classification no distinction between hyperconcentrated flows and clearwater floods is made. Therefore, watersheds where the propagation area does not extend to the mouth of the drainage area will be classified as clearwater flood dominated.

In order to assess the validity of the MSF algorithm, the debris propagation results were compared with the extent of a well-documented debris flow event occurred in the study area.

2.2.2.2 Development of the land cover indicator

The land cover indicator was constructed by analysing the characteristic land cover of each watershed, which was obtained from the classification of a LANDSAT TM5 image taken in 2001. The LANDSAT image was classified using a supervised classification algorithm. The reflectance values for different spectral wavelengths were extracted from the LANDSAT image for training samples with known land cover obtained from the inspection of a high-resolution Google image. The reflectance data of the training samples were used in a recursive partitioning algorithm from which a classification tree is obtained and applied to all pixels of the LANDSAT multiband image to establish separability of the classes based on the spectral signatures.

The classification identified areas covered by forests, grass, paramo vegetation [1], urban soil and water. From the land cover composition of each watershed a qualitative condition was derived.

The natural susceptibility of a catchment to debris flow hazards due to geological, morphological and climatic predispositions can be enhanced by human activities and the effects of land use changes [Koscielny et al., 2009]. In order to include this influence in the susceptibility analysis, the percentage of vegetation cover, urban area and bare soil were used to qualify the state of the watersheds.

Vegetation cover has been recognized as one of the factors related to frequency of debris flows [Jakob, 1996]. Forests reduce hydrogeomorphic hazards since they retain organic and inorganic material; contain the transport of mobilized material reducing the extent of destruction; intercept precipitation; and the stems of trees reduce the areas disturbed by snow avalanches, rockfalls, floods, debris floods and debris flows [Sakals et al., 2006]. Runoff can be increased by deforestation, soil properties

---

[1]Paramo is an alpine tundra ecosystem unique to the Andean Cordillera

degradation and impervious surfaces construction [Koscielny et al., 2009] as a result of urbanization. Likewise, erosion processes and slope instabilities can occur [Koscielny et al., 2009]. The percentage of bare soil represents areas prone to erosion and normally associated with quarries that can provide a supply of sediment.

According to Schueler [1995] stream degradation occurs at approximately 10-20% total impervious area. The increase in frequency and severity of floods due to imperviousness produces an increase in stream cross-sectional area. This occurs as a response of the stream accommodating higher flows through widening of the stream banks, downcutting of the stream bed or both. The channel instability triggers streambank erosion and habitat degradation. With respect to flood magnitude, this can be increased significantly by percentages of impervious cover larger than 10 percent. Hollis [1975] found that peak flows with recurrence intervals of 2-years increased by factors of two, three, and five with 10, 15 and 30 percent impervious development. A threshold of 15% was used to consider a high condition of urbanization of the watersheds and therefore a high degree of degradation. In order to consider the degree of degradation related to bare soil, normally related to quarries in the study area, a threshold of 10% was used.

2.2.2.3 Development of a composite susceptibility index

The resulting indicators of land cover and morphometry were combined using a matrix that allows classification of the catchments into high, medium and low susceptibility. Figure 2.3 shows the initial matrix used for the analysis. The corners corresponding to poor land cover and high morphometric indicator and good land cover and low morphometric indicator (cells a and f) were assigned a high and low susceptibility respectively, since they correspond to the extreme conditions in the analysis. The cells from b to h in Figure 2.3 were considered to potentially correspond to any category (low, medium or high priority) and all the possible combinations of the matrix were tested assessing the proportion correct of a contingency table comparing the obtained susceptibility index and the classification of flow type from the flood records, where debris flows were considered the most hazardous type of events. Potentially 2187 combinations can be obtained by assigning the three susceptibility categories to cells b to h in the matrix shown in Figure 2.3. Even if some combinations of the categories are not consistent with a progressive increase of susceptibility level from the bottom right corner of the matrix to the top left corner, all of them were tested. Under this procedure, the resulting matrix corresponds to the best fit of the susceptibility index and the classification of flow from flood records.

2.3 Results

The results obtained for each stage of the process are presented in the following subsections. The first subsection presents the results on the estimation of the morphometric indicator for the study area.

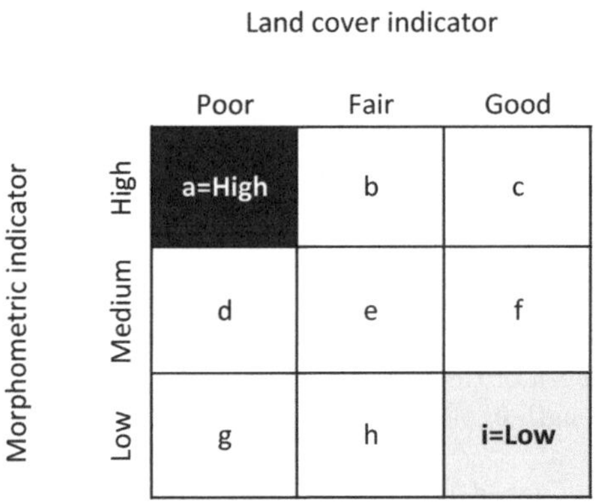

Land cover indicator

FIGURE 2.3: Matrix of classification of susceptibility

This subsection includes the development of the morphometric indicator model based on the principal component analysis and the assessment of the appropriateness of the morphometric indicator. The latter covers the classification of watersheds according to the debris flows propagation capacity and the comparison of the morphometric indicator with the propagation of debris flows described using the MSF model; with the 11 watersheds with confirmed flow type in the study area; and with three additional watersheds with confirmed flow type outside the study areas. The second subsection shows the results of the development of the land cover indicator and finally the third subsection shows the results of the combination of the morphometric indicator and the land cover indicator to obtain a final susceptibility index.

2.3.1 Estimation of the morphometric indicator for the study area

2.3.1.1 Morphometric indicator model

The results of the principal component analysis applying a varimax rotation carried out on the morphometric variables are shown in Table 2.2. From the Scree tests carried out on the eigen values obtained from the principal component analysis, the amount of principal components to be used were found to be 4. These first four principal components account for 85 percent of the variance in the data. From the analysis 4 groups of variables could be identified related to the size (inversely proportional to area), shape (proportional to circularity), hypsometry (proportional to hypsometric integral) and potential energy (proportional to the Melton number).

TABLE 2.2: Principal components and corresponding variables. The symbol column shows the abbreviation used in the formulas and Loading corresponds to the correlation of each variable with the corresponding principal component. Variables belonging to the PC1 were log transformed and variables with the symbol * were transformed as 1 - (value - minimum input value) / (maximum input value - minimum input value)

Variable	Symbol	Loading
PC1 - Size - % of Variability Explained=30%		
log)Perimeter*	P	0.96
log)Length of the watershed*	L_{wshd}	0.97
(log)Length of the main stream*	L_{Str}	0.95
(log)Area*	A	0.92
(log)Watershed Width*	W_{wshd}	0.83
PC2 - Shape - % of Variability Explained=28%		
Elongation ratio	E	0.93
Watershed legth to width*	LW	0.93
Circularity coeffiecient	C	0.95
Shape factor	SF	0.90
Drainage density*	DrD	0.66
PC3 - hypsometry - % of Variability Explained=22%		
Hypsometric skewness*	Hs	0.98
Hypsometric integral	Hi	0.90
Density skewness*	DHs	0.88
Hypsometric kurtosis*	Hk	0.91
Density kurtosis*	DHk	0.37
PC4 - Energy - % of Variability Explained=20%		
Relief ratio	R_{ra}	0.85
Watershed slope	S	0.89
Stream slope	StrS	0.63
Melton number	M	0.72
MRI mean*	MRIm	0.90

Using the factor loadings obtained from the first four principal components and scaling them to unity, the following equations were obtained:

$$P_{size} = 0.21L_{Str} + 0.22P + 0.20A + 0.22L_{wshd} + 0.16W_{wshd} \qquad (2.3)$$

$$P_{shape} = 0.21SF + 0.23C + 0.22E + 0.22LW + 0.11DrD \qquad (2.4)$$

$$P_{hypso} = 0.27Hs + 0.23Hi + 0.23Hk + 0.22DHs + 0.04DHk \qquad (2.5)$$

$$P_{energy} = 0.12StrS + 0.24S + 0.23R_{Ra} + 0.16M + 0.25MRIm \qquad (2.6)$$

The transformation of the variables in the analysis was made in such a way that the higher the value of the component the higher the flashiness or debris flow susceptibility. From the variability explained

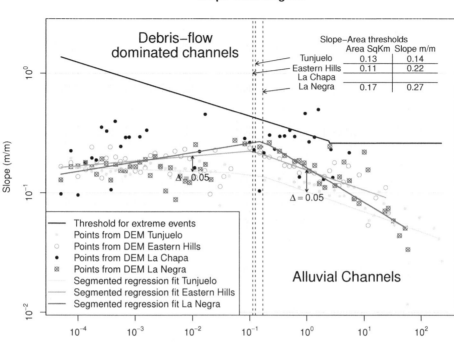

FIGURE 2.4: Slope-Area diagram for the study area and comparative areas. This figure shows the log slope versus log area for each pixel in the watershed areas. To increase readability the value of the slope is averaged in bins of 0.2 log of the drainage area. The black line corresponds to the curve of extreme events given by equation 2.1 and equation 2.2.

by each principal component, the morphometric indicator would be:

$$P_{morp} = 0.28P_{shape} + 0.20P_{hypso} + 0.22P_{energy} + 0.30P_{size} \qquad (2.7)$$

2.3.1.2 Assessment of appropriateness of the morphometric indicator

The slope–area relationship for the two regions of the study area (Tunjuelo basin and Eastern Hills) and the two external comparative watersheds (Negra creek and La Chapa creek) are shown in Figure 2.4.

For the Tunjuelo, Eastern Hills and La Negra watersheds the break in the slope-area diagram according to the segmented regression is located in a range of 0.11-0.17 km^2 for slopes between 0.14 and 0.27.

This is in the range of the values found by other authors for transition from debris flows to alluvial processes [Montgomery and Foufoula-Georgiou, 1993, Santos and Duarte, 2006, Seidl and Dietrich, 1993]. The points that belong to the La Chapa watershed do not allow the identification of a threshold. The drainage area of this watershed is only 7 km^2 which makes the identification of the break difficult (see Figure 2.4). Despite the significant scatter of the values, the slope–area points of La Chapa Creek are located above the points of the other watersheds (see Figure 2.4). Regarding the comparison of the points with the threshold of extreme events defined by Horton and Jaboyedoff [2008], the slope–area points of La Chapa watershed are located close and above the threshold for areas from 0.02 to 10 km^2. None of the other watersheds reach the threshold of extreme events, although, the points of the La Negra watershed are close to the threshold for areas between 2 and 10 km^2 (see Figure 2.4). The points of the Tunjuelo river basin, in general, lie lower than the points of the other watersheds and are thus more distant from the threshold of extreme events. It can be observed that the segmented regression of the Tunjuelo river basin is located below the segmented regression of the watersheds located in the Easter Hills, with a difference of approximately 0.05 m/m in slope (see Figure 2.4).

The propagation for initiation points that meet the slope-area thresholds was calculated using the MFS algorithm. However, this appears to overestimate the number of debris flow dominated watersheds. JICA [2006] identified slope failure areas related with debris flow occurrence in four watersheds located in the centre of the study area using aerial photographs from 1997 to 2004 (see Figure 2.5). The method applied by JICA [2006] identifies recent slope failures, old slope failures and mass movements related with potential debris flow initiation. In order to assess the initiation points obtained from the two approaches applied in this study, these were grouped into clusters where the distance between points is less than 50 m, in such a way that the clusters represent an area that produces the same propagation trajectory as the individual points.

The photointerpretation carried out by JICA [2006] resulted in 108 areas of failure. The slope-area threshold procedure, correctly identified 82% of these slope failure areas with 107 clusters lying on the areas identified by JICA [2006]. In contrast, the extreme event threshold correctly identified 65% of the slope failure areas with 103 clusters lying on the slope failure areas. Regarding the amount of initiation clusters identified by each criterion, the slope–area threshold resulted in 389 clusters, while the extreme events criteria identified 299 clusters. The slope–area threshold results in a false positive rate of 72%, and the extreme event threshold in a false positive rate of 66%. The visual comparison of the initiation points is shown in Figure 2.5. For the case of the slope-area threshold the clusters are scattered covering the mountainous area of the watersheds and even if they intersect the failure areas the clusters cover significant areas out of them, without showing a pattern associated to the past landslides. In the case of the initiation points from the extreme events threshold, these are not scattered on the upper watersheds but concentrated in areas from which 65% correspond to past failures. Even if the false positive rates for both methods are high, the overestimated amount and distribution of initiation points in the case of the slope–area threshold procedure leads to unrealistic results when the propagation is applied with propagation areas occupying most of the area of the

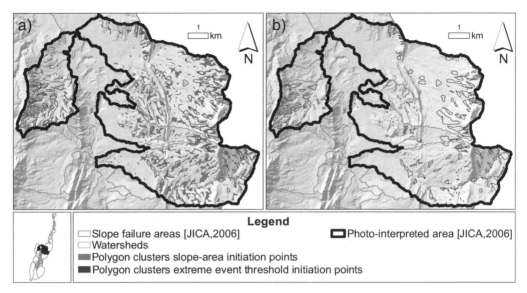

FIGURE 2.5: Comparison of failure areas detected by JICA [2006] and initiation points identified through a) the slope-area and b) the extreme event threshold

watersheds. Therefore, the propagation was recalculated using only the points above the curve of extreme events.

To assess the performance of the MSF algorithm, the propagation area was compared with the survey of the inundation extent of the debris flow occurred on the 19th of May 1994 in the upper basin of the Chiguaza Creek. This event affected 830 people and caused the death of 4 people [JICA, 2006]. Figure 2.6 shows the results of the propagation for the Chiguaza Creek. The inset shows the extent of the debris flow using the MSF algorithm as well as the observed extent. The comparison of the modelled and observed runout distance of this event shows a good agreement, although deviations from the modelled propagation areas exist in the final part of the runout (see inset). The deviations occur at bridges, which agrees with the analysis of the event carried out by JICA [2006] that concluded that obstructions in crossings had significantly influenced the trajectory of the flow. Simplified models like MSF cannot take the influence of bridges on the propagation of the flow into account. However, independent of the trajectory, the model seems to represent fairly well the downstream extent of the flow which is the main result needed for the analysis carried out in this study, since the distance between the simulated and observed downstream limit is only 60 meters.

Once the results of the MSF algorithm were obtained in the study area, these were used to classify the watersheds according to their capacity to propagate debris flows with two angles of reach. Figure 2.7 shows the distribution of the morphometric indicator against the classification of the watersheds according to the angle of reach. A clear differentiation can be seen for watersheds classified as 0.19H/L (able to propagate debris flows to their fans with an angle of reach of 0.19), with the lower quartile

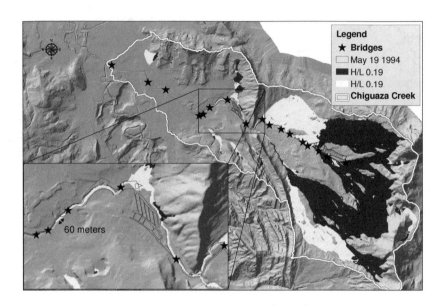

FIGURE 2.6: Affected area in the Chiguaza creek on 19th of May 1994 compared with propagation areas obtained from the MSF model

located above the interquartile ranges of the other two classifications. However, the differentiation between watersheds classified as 0.11H/L (able to propagate debris flows to their fans with an angle of reach of 0.11) and clear water watersheds (C) is less clear. Even if the lower and upper quartiles of the 0.11H/L watersheds are higher, the median value is smaller than the C watersheds. From this result, a qualitative subdivision into categories was made on the basis of the indicator. Low values from 0 to 0.35 correspond to watersheds unable to propagate debris flows to their fans according to the MSF algorithm, medium values from 0.35 to 0.61 correspond to watersheds where a propagation is possible with a reach angle of 0.11 and high values from 0.61 to 1 correspond to watersheds that can propagate debris flows with an angle of reach of 0.19.

Figure 2.8 shows the results of the morphometric indicator and its comparison with the results from the debris flow propagation algorithm. Figure 2.8-a shows the values of the morphometric indicator and its classification according to Figure 2.7, in this the flood records and the observed type of flow are overlaid. The definition of the observed flow type was possible for 11 watersheds, where the flood records, reports and available studies provide enough information to classify the watersheds into clear flow, hyperconcentrated flow and debris flow according to the method explained in section 2.2. Figure 2.8-b shows the resulting propagation areas for different angles of reach. The corresponding classification of the watersheds is shown in Figure 2.8-c depending on whether or not the lowest point of the watershed is reached by the propagation areas according to the angle of reach condition. The comparison of the spatial distribution of the morphometric indicator and the available flood records

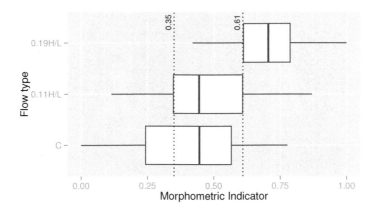

FIGURE 2.7: Morphometric indicator with values rescaled from 0 to 1

shows that the area of highest density of flood records is located in the centre of the study area, where the morphometric indicator ranges from 0 to 0.61 (see Figure 2.8-a).

A contingency table was constructed to assess the skill of the morphometric indicator to identify watersheds with the capacity to propagate debris flows to the fans according to the MSF model considering 0.11H/L watersheds less dangerous than 0.19H/L watersheds since the former are more fluid. The results are shown in Figure 2.9-a. When the three categories of the morphometric indicator are compared with the three flow classifications from the MSF model for all the watersheds in the study area, the proportion correct (PC) given as the fraction of the watersheds correctly identified is 0.56. When the contingency table is reduced to 2x2 dimensions, this is when only the identification of clear water and debris flow watersheds is assessed considering low values of the indicator associated to clear water flows and high and medium values associated to debris flows for angles of reach of 0.19 and 0.11, the proportion correct reaches 0.75.

The contingency table to assess the skill of the morphometric indicator to identify the observed flood types in the study area is shown in Figure 2.9-b. The 3x3 contingency table for the 11 watersheds for which flood type classification was possible, results in a value of 0.36 for the proportion correct, while the 2x2 contingency table provides a proportion correct of 0.55.

The values of the morphometric indicator obtained for the three watersheds outside of the analysed for the development of the morphometric indicator correspond to 0.3 in the case of the subwatershed of the Chiguaza Creek, 0.43 in the case of La Chapa Creek and 0.14 in the case of La Negra Creek.

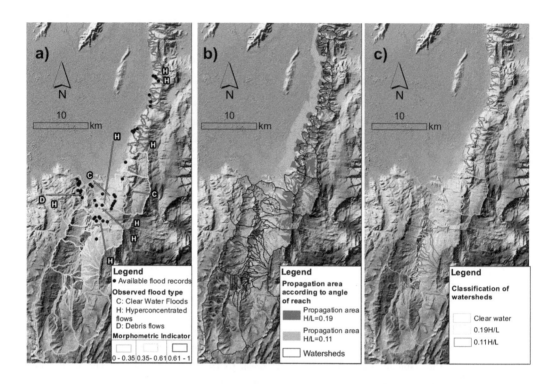

FIGURE 2.8: a) Morhometric indicator, b) propagation of debris flows, c) classification of watersheds

2.3.2 Land cover indicator

The three factors used to qualify the state of the watersheds (percentage of vegetation cover, percentage of urban area and percentage of bare soil) are shown in the ternary plot in Figure 2.10-a, where five areas were identified. As explained in section 2.2.2, limits for intensive degradation of the watershed were established taking into account the percentage of vegetation cover and bare soil cover (15% and 10% respectively) an additional limit was introduced in the ternary plot of 50% of vegetation cover that delimits area D in Figure 2.10-a, which represents watersheds with low urban use but high bare soil with low vegetation cover. Watersheds corresponding to zones C, D and E in Figure 2.10-a were grouped into watersheds in poor condition, watersheds in zone B correspond to fair condition and watersheds in zone A to good condition. The position of the dots in the ternary plot represents the conditions of the watersheds of the study area. Most of the dots are located in zone A. However, highly urbanized watersheds with poor vegetation cover and bare soil can be identified. The spatial distribution of these watersheds can be observed in Figure 2.10-b where a critical area can be localized in the lower part of the Tunjuelo river basin.

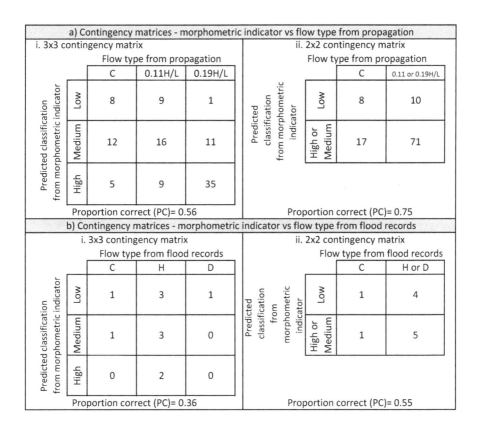

FIGURE 2.9: Contingency table to compare the watershed classification according to debris flow propagation capacity from the MSF model and the morphometric indicator; and the flood type classification from available information and the morphometric indicator.

2.3.3 Combination of indicators to obtain a final susceptibility index

From the tests on the possible combination matrices defined by the structure showed in Figure 2.3, the highest proportion correct that was obtained was 0.75 considering the three susceptibility classifications and the three types of flow obtained for the 11 watersheds where information was enough to carry out the classification. The debris flows were assigned the most dangerous condition. A proportion correct of 0.91 was obtained when only distinction between clear water flows and debris flows was considered. The optimum matrix is shown in Figure 2.11-a and Figure 2.11-b shows the contingency matrices.

Figure 2.12 shows the resulting classification of the watersheds applying the matrix shown in Figure 2.11-a. In this, observed occurrence of floods was superimposed on the susceptibility classification

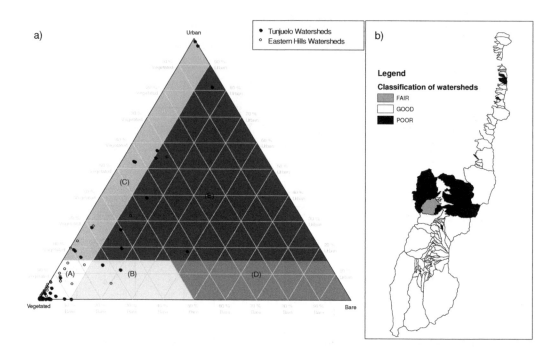

FIGURE 2.10: a) Ternary plot for classification of watersheds according to landcover. The description of the zones of the plot is as follows: (A) low percentage bare soil, low percentage of urban soil and high percentage of vegetated areas; (B) high percentage of bare soil, low percentage of urban soil and high percentage of vegetated areas; (C) low percentage of bare soil, high percentage of urban land and low percentage of vegetated areas; (D) high percentage of bare soil, low percentage or urban soil and low percentage of vegetated land; (D) high percentage of bare soil, high percentage of urban area and low percentage of vegetated cover. b) Classification of watersheds according to landcover

where each dot represents a recorded flood. The spatial distribution of the flood events clearly concentrates in the watersheds located in the lower basin of the Tunjuelo river where there is a cluster of watersheds classified as medium and high susceptibility.

2.4   Discussion

2.4.1   Morphometric indicator

Figure 2.13 shows the boxplots of the composite morphometric indicator and the individual indicators for size, energy, hypsometry and shape. The indicators were grouped according to the classification

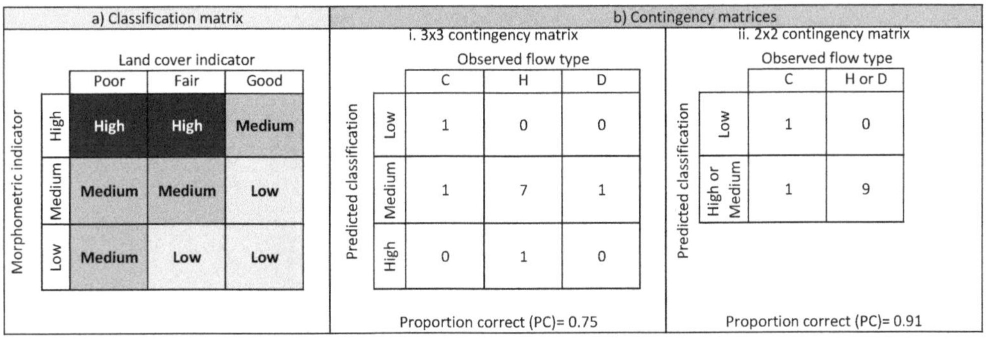

a) Classification matrix				b) Contingency matrices							

FIGURE 2.11: b) Optimum classification matrix, b) Contingency table to compare the watershed classification according to the composite indicator (morphometric indicator and land cover indicator) and the observed flow type

of the watersheds carried out on the basis of the capacity to propagate debris flows to the fan of the watershed. The indicators calculated for La Negra Creek watershed, La Chapa Creek watershed and the subwatershed of Chiguaza Creek (drainage area to the most downstream point affected by the debris flow on May 19 1994) were plotted over the boxplots. La Chapa creek was classified as 0.19H/L, and Negra creek and the subwatershed of Chiguaza creek as 0.11H/L according to the results of the MSF algorithm applied in these watersheds.

The comparison of the composite morphometric indicator of the watersheds in the study area with that of the Chiguaza, La Chapa and La Negra watersheds, shows that the latter watersheds have a low indicator. This is mainly due to the size indicator, that in comparison with the size of the watersheds in the areas assigns low values, with the lowest being the indicator of La Negra Creek which has an area of 68.4 km^2 (the largest area in the analysis). It is important to take into account that the composite morphometric indicator not only involves the capacity of the watershed to propagate debris flows but also the flashiness, this means that watersheds with the characteristics to propagate debris flows are not necessarily the flashiest.

From the results of the size indicator shown in Figure 2.13-b, it can be observed that in general watersheds classified as 0.19H/L exhibit high values of the indicator. However, the size indicator does not discriminate between processes. This can be due to the scale of the analysis, since all the analysed watersheds can be considered small.

In the principal component analysis, the Size Indicator has the highest weight in the total morphometric indicator (see Equation 2.7). Several studies have shown that drainage area is correlated with other morphometric parameters, for example, De Scally and Owens [2004] suggest that drainage area acts as a surrogate for the channel gradient and Gray [1961], and Shreve [1974] showed the correlation

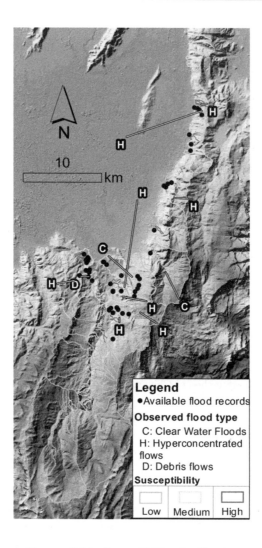

FIGURE 2.12: Susceptibility classification

between length of the main stream and drainage area. Similar to the findings of other authors [Gray, 1961, Mesa, 1987, Shreve, 1974] a high determination coefficient was found between the logarithm of the stream length and the logarithm of drainage area (R^2=0.92). The same behaviour is exhibited by the logarithm of length of the watershed (R^2=0.90), logarithm of watershed width (R^2=0.95) and logarithm of perimeter of the watershed (R^2=0.96). The empirical relationship between length of the main stream (longest stream) and the area is known as Hack's law [Hack, 1957]. The exponent of the power law may vary slightly from region to region, but it is generally accepted to be slightly below 0.6 [Rigon et al., 1996]. In the study area this exponent corresponds to 0.59. Several authors have tried to explain the relationship between main stream length and basin area [Mantilla et al.,

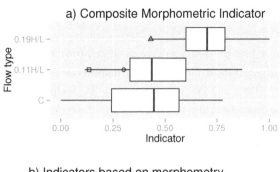

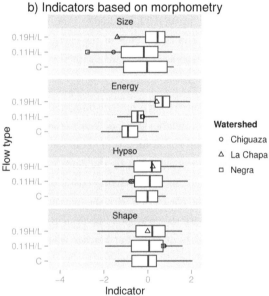

FIGURE 2.13: a) Composite Morphometric Indicator, b) Indicators based on morphometry. NOTE: 0.19H/L and 0.11H/L correspond to watersheds that can propagate debris flows to their fans considering angles of reach of 0.19 and 0.11 respectively.

2000]. The conclusion reached by Willemin [2000] indicates that there is some aspect of the evolution of fluvial systems not yet understood, that somehow takes into account three geometric components (basin elongation, basin convexity and stream sinuosity), none of which is particularly well correlated with basin area, and produces a robust relationship between main stream length and basin area. This conclusion is coherent with the findings of this study, where elongation does not show a strong correlation with basin area nor a trend to more elongated basins with increasing size of the watershed, as it is not in the same principal component (see Table 2.2).

The energy indicator, which provides a measure of the potential energy, is composed of the relief ratio, the mean watershed slope, the stream slope, the Melton number and the mean of the multiresolution index (MRI). As suggested by Gallant and Dowling [2003], the MRI can lead to identify similarities and differences between catchments which in this analysis correspond to the energy of the watersheds. High Melton numbers have been previously used as an effective discriminator of debris flow dominated watersheds. However, the threshold for the Melton number varies significantly depending on the region, ranging from 0.5 [Welsh and Davies, 2010] to 0.75 [De Scally and Owens, 2004]. Despite the variability of its components, the energy indicator clearly distinguishes 0.19H/L watersheds. Some superposition of values occurs but the interquartile range of 0.19H/L is separated from the interquartile ranges of the other two classifications (see Figure 2.7). In terms of energy it is more difficult to distinguish between 0.11H/L watersheds and C watersheds. However, the mean and the first and third quartiles of the energy indicator for 0.11H/L are higher than in the case of C watersheds, but with a wider range of superposition. The high values of the energy indicator for the subwatershed of Chiguaza, La Chapa and La Negra creeks is consistent with the processes that take place in the watersheds.

Regarding the hypsometric indicator, since the hypsometric integral decreases as mass is removed from the watershed it follows that an inverse relationship between hypsometric skewness and the hypsometric integral exists [Harlin, 1984]. This condition was found in the study area with a determination coefficient of 0.71. The same behaviour is exhibited by the density skewness (R^2=0.82) and hypsometric Kurtosis (R^2=0.45), where small values are characteristic of large integral values and small skewness. The density kurtosis shows no correlation with the hypsometric integral, this is reflected in the low correlation of this parameter with the corresponding principal component in the analysis (see Table 2.2). Headward erosion that starts at the lower reaches would represent a higher possibility of debris flow affecting the urbanized fans of the watersheds; therefore this increase in susceptibility would be represented by high hypsometric integrals, low hypsometric skewness and negative density skewness. Furthermore, according to Cohen et al. [2008] higher hypsometric integral values (greater than 0.5) represent catchments dominated by diffusive erosion processes (concave down hypsometric curve) while lower values (less than 0.5) represent fluvial dominated catchments (concave up hypsometric curve). Therefore the hypsometric integral is linked to erosion processes, landform curvature and landscape morphology.

The boxplots of the hypsometric indicator (see Figure 2.13-b) do not show a differentiation according to the classification of watersheds based on the capacity to propagate debris flows. The superposition of the values of the hypsometric indicator obtained for Chiguaza, La Chapa and Negra creeks shows interesting results, mainly for the case of La Chapa creek where the value can be classified as high in comparison with the other watersheds. Linking this result with the slope–area plot (Figure 2.4) where no fluvial dominated area was identified for La Chapa creek, it can be inferred that there is a dominance of diffusive processes (characteristic of debris flows) in this watershed that is captured by the morphometric indicators. Therefore, the hypsometric indicator may contribute to explain the dominance of processes that supply sediment in the watershed. The availability of sediment is one of the determining factors for the occurrence of debris flows. However, its assessment requires extensive field work and detailed sediment source analysis. This assessment is not replaced by the hypsometric indicator, but for the scale of the analysis this indicator is considered to significantly contribute in the susceptibility recognition.

For the case of the shape indicator, drainage density was found to be correlated with the principal component related to the shape of the watersheds, which confirms its relation with the physiographic characteristics of the watersheds [Gregory and Walling, 1968]. The boxplots of Figure 2.13-b shows that the capacity to transport debris flows is independent of the shape indicator. However, the values of the indicator for the three watersheds used as external test areas (Chiguaza, La Chapa and Negra) are in the range of high values, particularly Chiguaza and Negra creeks show a very high value. These two watersheds are very similar in terms of shape, hypsometry and energy.

High values of the indicator involve small area high energy watersheds with shapes that contribute to flashiness and hypsometric characteristics that imply erosive processes.

2.4.2   Debris flow propagation

The analysis of the slope-area curves shows that on average, the slope in La Chapa watershed is higher for a given drainage area than for the other watersheds considered. If the same drainage area, e.g. $1km^2$, is considered for the three watersheds with segmented regression fit shown in Figure 2.4, namely Tunjuelo river basin, Eastern Hills and La Negra creek, the slope values from the slope-area curves are 0.1, 0.15 and 0.16 respectively, which means that on average for this drainage area the local slope in the Tunjuelo river basin is milder than in the Eastern Hills with the latter being slightly milder than the local slope in La Negra creek. In the case of La Chapa watershed the value of slope for a drainage area of $1km^2$ is 0.4. This result is important given that La Chapa creek has a confirmed debris flow dominance, followed by La Negra creek where concentrations in the transition from hyperconcentrated flows and debris flows have been identified. High values of the morphometric indicator are concentrated in the watersheds located in the north east of the study area. This behaviour is in agreement with the characteristics of the slope-area diagram shown in Figure 2.4, where on average the watersheds in the Eastern Hills have higher local slope for a given area than in the Tunjuelo Basin watersheds. This condition reflects a difference in energy between the two areas that is captured by the morphometric indicator.

The differences in the threshold of extreme events and the location of the slope–area points belonging to each watershed imply that applying the threshold of extreme events reduces the amount of initiation points in comparison with the use of the threshold obtained from the slope–area relationship for dominance of debris flow processes. The comparison of initiation points obtained from the slope–area and from the threshold of extreme events with the failure areas obtained from photo interpretation, shows that the slope–area and initiation points seem to overestimate the amount of initiation points in the study area. It is important to highlight that the points correspond to values of local slope averaged in a range of area, therefore, even in the case of the Tunjuelo river basin individual points that meet the extreme event criteria can be identified. However, the amount is less than in the case of the other watersheds.

The results of the MSF algorithm using the threshold of extreme events show that independently of the classification of flow type of the watershed based on the type of flow at the mouth, other types of flow can occur in other areas of the watersheds, as is the case with the Chiguaza Creek where the extent of the propagation was compared with a field survey providing a good correspondence between the two. The classification of the type of flows at the mouth of this watershed is clearwater flow. However in upper areas were the supply of sediment is high, the morphometric conditions favour debris flows and the land cover is characterised by areas with bare soil.

2.4.3   Land cover indicator, composite susceptibility index and comparison of results

Even if the morphometric indicator provides insight in the expected behaviour and dominant processes of the watersheds reflecting the propagation capacity of the watersheds with a proportion correct of 0.56, it does not fully explain the distribution, characteristics and occurrence of the flood events in the study area. The proportion correct of the contingency matrix comparing the classification obtained from the morphometric indicator and the flow type from flood records yields a value of only 0.36.

When the land cover indicator was included in the analysis on the basis that the land cover can exert a positive influence in the case of vegetated surfaces, but also can enhance the susceptibility conditions when urban and bare soil areas are significant, the proportion correct of the contingency matrix comparing the resulting susceptibility indicator and the flow type from flood records increased to 0.75.

It is important to consider that the mountains of Bogotá, mainly in the south of the city and in some localized areas of the east, have been subjected to illegal urbanization processes. The processes involved in informal settlement entail the construction of houses in the creeks, in some cases not only in the protection buffers but also in the channels. Furthermore, urbanization requires river crossings that in many cases are not technically designed and constitute dangerous obstructions to the flow as presented in section 3.1.2. Another important aspect to consider is the accumulation of waste material in the channels, which during flood events is transported by the flow and obstructions are common in highly urbanized watersheds in the study area.

The inclusion of the land cover influence in the analysis helps to explain the highly deteriorated conditions of some of the watersheds located in the south of the city where floods are frequent, but also to explain the lower occurrence of flood events in some watersheds in the east of the city where the presence of forests and protected areas has contributed to preserve the natural conditions of the watersheds. This suggests the importance of taking land cover into account when assessing the susceptibility to different types of flash floods in peri-urban areas of cities in mountainous areas.

2.5 Conclusions

A susceptibility indicator composed of a morphometric indicator and a land cover indicator was used to classify the flash flood susceptibility of 106 watersheds located in the mountainous peri-urban areas of Bogotá (Colombia). Morphological variables recognized in literature to have a significant influence in flashiness and occurrence of debris flows were used to construct the morphometric indicator. Subsequently, this indicator was compared with the results of simplified debris flows propagation techniques; with the flood type classification carried out in 11 watersheds of the study area; and assessed in three additional watersheds to those analysed in the development of the morphometric indicator. These comparisons were made in order to assess the appropriateness of the morphometric indicator. A susceptibility index for each of the catchments was subsequently obtained through the combination of the morphometric indicator and a land cover indicator. An important consideration during the analysis is that watersheds that are prone to debris flows are more dangerous than other flashy watersheds.

The derived susceptibility index is not absolute, but relative, and is useful in applications at regional scales for preliminary assessment and prioritization of more detailed studies. A limitation of the method is that it does not take sediment availability into account, which is a determining factor for debris flow occurrence. Even if some morphometric indicators could be related to erosion and sediment availability, this factor should be assessed through other techniques.

The morphological variables that were identified to enhance debris flow hazard, were analysed through principal component analysis, finding that the 20 variables could be summarized in 4 component indicators related to size, shape, hypsometry and energy of the watersheds. Size of the watersheds is the component that has the highest weight in the development of the final morphometric indicator. This result is in agreement with previous research that identifies this parameter as relevant in the identification of hazard.

The use of the slope-area curve to identify debris flows source areas showed an overestimation of potential sources when compared with other methods using empirical thresholds. However, it provides valuable information on the processes occurring in a watershed. The slope-area diagram obtained regionally can provide insight in the susceptibility at morphometric level when curves are compared between watersheds in different areas. In the case of the study area, the comparison of the slope-area curves of the Tunjuelo basin and the Eastern hills watersheds, allowed to conclude that the latter exhibit on average a higher slope for a given area, which is reflected in the energy indicator that is linked to the capacity to transport debris flows.

The energy indicator was shown to distinguish watersheds with the capacity to transport debris flows to their fans. This indicator involves parameters previously successfully used to identify debris flow dominated watersheds. While the prevalence of debris flows in a watershed should be confirmed using detailed information on geology and geotechnics, this parameter can be taken as an initial assessment and for prioritization where to focus such detailed studies.

The use of size, shape and hypsometry indicators in addition to the energy indicator, contribute to include valuable information in the analysis to integrally assess the watersheds. Size includes information regarding flashiness as well as shape. Hypsometry was found to be a promising indicator regarding the geomorphic evolution of the watershed and the erosion.

Despite the ability of the morphometric indicator to identify the capability to transport debris flows, it was found not to be sufficient to explain the records of past floods in the study area. The land cover indicator was included, with the objective to involve in the analysis not only the benefit of vegetated areas but also the enhancement of hazard conditions produced by urbanization and soil deterioration. The indicator produced by the combination of the morphometric indicator and the land cover indicator improved the agreement between the results of the classification and the records of past floods in the area. This implies that even if morphometric parameters show a high disposition for debris flow, land cover can compensate and reduce the susceptibility. On the contrary, if favourable morphometric conditions are present but deterioration of the watershed takes place the danger increases.

Chapter 3

Regional prioritisation of flood risk in mountainous areas

This chapter is an edited version of: Rogelis, M. C., Werner, M., Obregón, N., and Wright, N.: Regional prioritisation of flood risk in mountainous areas, Nat. Hazards Earth Syst. Sci., 16, 833-853, doi:10.5194/nhess-16-833-2016, 2016.

3.1 Introduction

Flood risk represents the probability of negative consequences due to floods and emerges from the convolution of flood hazard and flood vulnerability [Schanze et al., 2006]. Assessing flood risk can be carried out at national, regional or local level [IWR, 2011], with the regional scale aiming at contributing to regional flood risk management policy and planning. Approaches used to assess flood risk vary widely. These include the assessment of hazard using model-based hazard analyses and combining these with damage estimations to derive a representation of risk [Liu et al., 2014, Su and Kang, 2005], as well as indicator-based analyses that focus on the assessment of vulnerability through composite indices [Chen et al., 2014, Greiving, Stefan, 2006, Safaripour et al., 2012]. The resulting levels of risk obtained may subsequently be used to obtain grades of the risk categories (e.g. high, medium and low) that allow prioritisation, or ranking of areas for implementation of flood risk reduction measures, such as flood warning systems and guiding preparations for disaster prevention and response [Chen et al., 2014].

A risk analysis consists of an assessment of the hazard as well as an analysis of the elements at risk. These two aspects are linked via damage functions or loss models, which quantitatively describe how hazard characteristics affect specific elements at risk. This kind of damage or loss modelling, typically provides an estimate of the expected monetary losses [Luna et al., 2014, Mazzorana et al., 2012, Seifert et al., 2009, Van Westen et al., 2014]. However, more holistic approaches go further, incorporating social, economic, cultural, institutional and educational aspects, and their interdependence [Fuchs, 2009]. In most cases these are the underlying causes of the potential physical damage [Birkmann et al., 2014, Cardona, 2003, Cardona et al., 2012]. A holistic approach provides crucial information that supplements flood risk assessments, informing decision makers on the particular causes of significant losses from a given vulnerable group and providing tools to improve the social capacities of flood victims [Nkwunonwo et al., 2015]. The need to include social, economic and environmental factors, as well as physical in vulnerability assessments, is incorporated in the Hyogo Framework for Action and emphasized in the Sendai Framework for Disaster Risk Reduction 2015-2030, which establishes as a priority the need to understand disaster risks in all its dimensions [United Nations General Assembly, 2015]. However, the multi-dimensional nature of vulnerability has been addressed by few studies [Papathoma-Köhle et al., 2011].

The quantification of the physical dimension of vulnerability can be carried out through empirical and analytical methods [Sterlacchini et al., 2014]. However, when the multiple dimensions of vulnerability are taken into account, challenges arise in the measurement of aspects of vulnerability that can not be easily quantified. Birkmann [2006] suggests that indicators and indices can be used to measure vulnerability from a comprehensive and multidisciplinary perspective, capturing both direct physical impacts (exposure and susceptibility), and indirect impacts (socio-economic fragility and lack of resilience). The importance of indicators is rooted in their potential use for risk management since they are useful tools for: (i) identifying and monitoring vulnerability over time and space; (ii) developing an improved understanding of the processes underlying vulnerability, (iii) developing and prioritising strategies to reduce vulnerability; and for (iv) determining the effectiveness of those strategies [Rygel et al., 2006]. However, developing, testing and implementing indicators to capture the complexity of vulnerability remains a challenge.

The use of indices for vulnerability assessment has been adopted by several authors, for example, Balica et al. [2012] describe the use of a Flood Vulnerability Index, an indicator-based methodology that aims to identify hotspots related to flood events in different regions of the world. Müller et al. [2011] used indicators derived from geodata and census data to analyse the vulnerability to floods in a dense urban setting in Chile. A similar approach was followed by Barroca et al. [2006], organising the choice of vulnerability indicators and the integration from the point of view of various stakeholders into a software tool. Cutter et al. [2003] constructed an index of social vulnerability to environmental hazards at county-level for the United States. However, several aspects of the development of these indicators continue to demand research efforts, including: the selection of appropriate variables that are capable of representing the sources of vulnerability in the specific study area; the determination of the importance of each indicator; the availability of data to analyse and assess the indicators; the limitations in the scale of the analysis (geographic unit and timeframe); and the validation of the results [Müller et al., 2011]. Since, no variable has yet been identified against which to fully validate vulnerability indicators, an alternative approach to assess the robustness of indices is to identify the sensitivity of how changes in the construction of the index may lead to changes in the outcome [Schmidtlein et al., 2008].

Vulnerability is closely tied to natural and man made environmental degradation at urban and rural levels [Cardona, 2003, UNEP, 2003]. At the same time the intensity or recurrence of flood hazard events can be partly determined by environmental degradation and human intervention in natural ecosystems [Cardona et al., 2012]. This implies that human actions on the environment determine the construction of risk, influencing the exposure and vulnerability as well as enhancing or reducing hazard. For example, the construction of a bridge can increase flood hazard upstream by narrowing the width of the channel, increasing the resistance to flow and therefore resulting in higher water levels that may inundate a larger area upstream.

The interaction between flood hazard and vulnerability is explored in small watersheds in a mountainous environment, where human-environment interactions that influence risk levels take place in a limited area. The hydrological response of these watersheds is sensitive to anthropogenic interventions, such as land use change [Seethapathi et al., 2008].

The consequence of the interaction between hazard and vulnerability in such small watersheds is that those at risk of flooding themselves play a crucial role in the processes that enhance hazard, through modification of the natural environment. Unplanned urbanization, characterized by a lack of adequate infrastructure and socioeconomic issues (both contributors to vulnerability) may also result in environmental degradation, which increases the intensity of natural hazards [United Nations and ISDR, 2004]. In the case of floods, such environmental degradation may lead to an increase in peak discharges, flood frequency and sediment load.

In this study a method to identify mountain watersheds with the highest flood damage potential at the regional level is proposed. Through this, the watersheds to be subjected to more detailed risk studies can be prioritised in order to establish appropriate flood risk management strategies. The method is demonstrated in the mountain watersheds that surround the city of Bogotá (Colombia), where floods typically occur as flash floods and debris flows.

The prioritisation is carried out through an index composed of a qualitative indicator of vulnerability and a qualitative indicator of the susceptibility of the watersheds to the occurrence of flash floods/debris flows. Vulnerability is assessed through application of an indicator system that considers social, economic and physical aspects that are derived from the available data in the study area. This is subsequently combined with an indicator of flash flood/debris flow susceptibility that is based on morphometry and land cover, and was applied to the same area in a previous study [Rogelis and Werner, 2013]. In the context of the flash flood/debris flow susceptibility indicator, susceptibility is considered as the spatial component of the hazard assessment, showing the different likelihoods that flash floods and debris flow occur in the watersheds. In contrast, risk is defined as the combination of the probability of an event and its negative consequences [UNISDR, 2009]. The priority index can be considered a proxy for risk, identifying potential for negative consequences but not including probability estimations.

The chapter is structured as follows: (i) Section 2 reviews the conceptual definition of vulnerability as the foundation of the study; (ii) Section 3 describes the study area, and the data and methodology used; (iii) Section 4 presents the results of the analysis. This includes the construction of the indicators and the corresponding sensitivity analysis, as well as the prioritisation of watersheds; (iv) Section 5 interprets the results that lead to the final prioritisation; (v) The conclusions are summarised in Section 6.

3.2 Conceptualization of Vulnerability

Several concepts of vulnerability can be identified, and there is not a universal definition of this term [Birkmann, 2006, Thieken et al., 2006]. Birkmann [2006] distinguishes at least six different schools of thinking regarding the conceptual and analytical frameworks on how to systematise vulnerability. In these, the concept of exposure and its relation with vulnerability, the inclusion of the coping capacity as part of vulnerability, the differentiation between hazard dependent and hazard independent characteristics of vulnerability play an important role. [Sterlacchini et al., 2014] identifies at least two different perspectives: (i) one related to an engineering and natural science overview; and (ii) a second one related to a social science approach.

With relation to the first perspective (i), vulnerability is defined as the expected degree of loss for an element at risk, occurring due to the impact of a defined hazardous event [Fuchs, 2009, Holub et al., 2012, Varnes, 1984]. The relationship between impact intensity and degree of loss is commonly expressed in terms of a vulnerability curve or vulnerability function [Totschnig and Fuchs, 2013], although also semi-quantitative and qualitative methods exist [Fuchs et al., 2007, Jakob et al., 2012, Kappes et al., 2012, Totschnig and Fuchs, 2013]. The intensity criteria of torrent (steep stream) processes, encompassing clear water, hyperconcentrated and debris flows, has been considered in terms of impact forces [Holub et al., 2012, Hu et al., 2012, Quan Luna et al., 2011]; deposit height [Akbas et al., 2009, Fuchs et al., 2007, 2012, Lo et al., 2012, Mazzorana et al., 2012, Papathoma-Köhle et al., 2012, Totschnig and Fuchs, 2013, Totschnig et al., 2011]; kinematic viscosity [Quan Luna et al., 2011, Totschnig et al., 2011], flow depth [Jakob et al., 2013, Totschnig and Fuchs, 2013, Tsao et al., 2010]; flow velocity times flow depth [Totschnig and Fuchs, 2013]; and velocity squared times flow depth [Jakob et al., 2012]. Different types of elements at risk will show different levels of damage given the same intensity of hazard [Albano et al., 2014, Jha et al., 2012, Liu et al., 2014], therefore vulnerability curves are developed for a particular type of exposed element (such as construction type, building dimensions or road access conditions). A limited number of vulnerability curves for torrent processes have been proposed, and the efforts have been mainly oriented to residential buildings [Totschnig and Fuchs, 2013]. Since it can be difficult to extrapolate data gathered from place to place to different building types and contents [Papathoma-Köhle et al., 2011], different curves should be created for different geographical areas and then applied to limited and relatively homogeneous regions [Fuchs et al., 2007, Jonkman et al., 2008, Luino et al., 2009] .

Regarding the second perspective (ii), social sciences define vulnerability as the pre-event, inherent characteristics or qualities of social systems that create the potential for harm [Cutter et al., 2008]. This definition is focused on the characteristics of a person or group and their situation than influence their capacity to anticipate, cope with, resist and recover from the impact of a hazard [Wisner et al., 2003]. Social and place inequalities are recognized as influencing vulnerability [Cutter et al., 2003]. The term livelihood is highlighted and used to develop models of access to resources, like money, information, cultural inheritance or social networks, influencing people's vulnerability [Hufschmidt et al., 2005].

Given the different perspectives of vulnerability it becomes apparent that only by a multidimensional approach, the overall aim of reducing natural hazards risk can be achieved [Fuchs and Holub, 2012]. Fuchs [2009] identifies a structural (physical) dimension of vulnerability that is complemented by economic, institutional and societal dimensions. In addition to these, Sterlacchini et al. [2014] identify a political dimension. Birkmann et al. [2014] and Birkmann et al. [2013] identify exposure, fragility and lack of resilience as key causal factors of vulnerability, as well as physical, social, ecological, economic, cultural and institutional dimensions.

In this study, physical exposure (hard risk and considered to be hazard dependent), socioeconomic fragility (soft risk and considered to be not hazard dependent) and lack of resilience and coping capacity (soft risk and is mainly not hazard dependent) [Cardona, 2001] are used to group the variables that determine vulnerability in the study area. In this study, the risk perception and the existence of a flood early warning, which are hazard dependent, are considered as aspects influencing resilience since they influence the hazard knowledge of the communities at risk and the level of organization to cope with floods. An analysis of physical vulnerability through vulnerability curves is not incorporated, instead the expected degree of loss is assessed qualitatively through the consideration of physical exposure and factors that amplify the loss (socioeconomic fragility and lack of resilience). This means the expected degree of loss depends on the extent of the flash floods/debris flows, and not on the intensity of those events.

The terminology and definitions that are used in this study are as follows:

- Vulnerability: propensity of exposed elements such as physical or capital assets, as well as human beings and their livelihoods, to experience harm and suffer damage and loss when impacted by a single or compound hazard events [Birkmann et al., 2014].

- Exposure: people, property, systems, or other elements present in hazard zones that are thereby subject to potential losses [UNISDR, 2009].

- Fragility: predisposition of elements at risk to suffer harm [Birkmann et al., 2014].

- Lack of resilience and coping capacity: limited capacities to cope or to recover in the face of adverse consequences [Birkmann et al., 2014].

3.3 Methods and Data

3.3.1 Study Area

Bogotá is the capital city of Colombia with 7 million inhabitants and an urban area of approximately 385 km^2. The city is located on a plateau at an elevation of 2640 meters above sea level and is surrounded by mountains from where several creeks drain to the Tunjuelo, Fucha and Juan Amarillo rivers. These rivers flow towards the Bogotá River. Precipitation in the city is characterised by a bimodal regime with mean annual precipitation ranging from 600 mm to 1200 mm [Bernal et al., 2007].

Despite its economic output and growing character as a global city, Bogotá suffers from social and economic inequalities, lack of affordable housing, and overcrowding. Statistics indicate that there has been a significant growth in the population, which also demonstrates the process of urban immigration that the whole country is suffering not only due to industrialization processes, but also due to violence and poverty. This disorganised urbanisation process has pushed informal settlers to build their homes in highly unstable zones and areas that can be subjected to inundation. Eighteen percent of the urban area has been occupied by informal constructions, housing almost 1,400,000 persons. This is some 22% of the urban population of Bogotá [Pacific Disaster Center, 2006].

Between 1951 and 1982, the lower (northern) part of the Tunjuelo basin (see Figure 3.1) was the most important area for urban development in the city, being settled by the poorest population of Bogotá [Osorio, 2007]. This growth has been characterised by informality and lack of planning. This change in the land use caused loss of vegetation and erosion, which enhanced flood hazard [Osorio, 2007].

The urban development of the watersheds located in the hills to the east of Bogotá (see Figure 3.1) has a different characteristic to that of the Tunjuelo basin. Not only has this taken place through both informal settlements, but also includes exclusive residential developments [Buendía, 2013]. In addition, protected forests cover most of the upper watersheds.

In this analysis the watersheds located in mountainous terrain that drain into the main stream of the Tunjuelo basin, as well as the watersheds in the Eastern Hills were considered. The remaining part

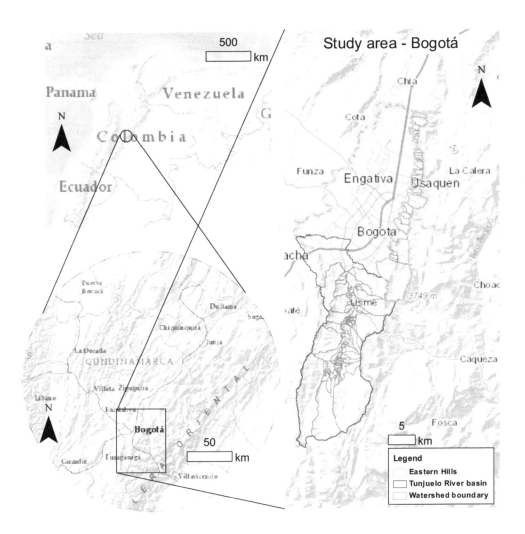

FIGURE 3.1: Location of the study areas. Service Layer Credits: Sources: Esri, HERE, DeLorme, Intermap, increment P Corp., GEBCO, USGS, FAO, NPS, NRCAN, GeoBase, IGN, Kadaster NL, Ordnance Survey, Esri Japan, METI, Esri China (Hong Kong), swisstopo, MapmyIndia, OpenStreetMap contributors, and the GIS User Community

of the urban area of the city covers an area that is predominantly flat, and is not considered in this study. Table 3.1 shows the number of watersheds in the study area, as well as the most recent and severe flood events that have been recorded.

TABLE 3.1: Most severe recent flooding events in the study area

Watersheds				
Study Area	Number	Average Slope (%)	Area (km^2)	Recent flooding events
Tunjuelo River Basin	66	12-40	0.2-57	The most severe events include: • In May 1994 a debris flow affected 830 people and caused the death of 4 people in the north east of the basin [JICA, 2006]. • In November 2003 a hyperconcentrated flow took place in the north west of the Tunjuelo basin. 2 people were killed and 1535 were affected. A similar event occurred at the same location in November/2004 without death toll [DPAE, 2003a,b].
Eastern Hills	40	21-59	0.2-33	The most sever events include: • In May 2005 a hyperconcentrated flow occurred in the central part of the area affecting 2 houses [DPAE, 2005].

3.3.2 Methodology

The prioritisation of flood risk was carried out using watersheds in the study area as units of analysis. The watershed divides were delineated up to the confluence with the Tunjuelo River, or up to the confluence with the storm water system, whichever is applicable. First a delineation of areas exposed to flooding from these watersheds using simplified approaches was carried out. Subsequently a vulnerability indicator was constructed based on a principal component analysis of variables identified in the literature as contributing to vulnerability. A sensitivity analysis was undertaken to test the robustness of the vulnerability indicator. From the vulnerability indicator a category (high, medium and low vulnerability) was obtained that was then combined with a categorisation of flash flood/debris flow susceptibility previously generated in the study area to obtain a prioritisation category. The tool that was used to combine vulnerability and susceptibility was a matrix that relates the susceptibility levels and vulnerability levels producing as output a priority level. The combination matrix was constructed through the assessment of all possible matrices using as assessment criterion the "proportion correct". In order to obtain the "proportion correct" an independent classification of the watersheds was carried out on the basis of the existing damage data.

A detailed explanation of the analysis is given in the following subsections.

3.3.2.1   Delineation of exposure areas

Flood events in the watersheds considered in this study typically occur as flash floods given their size and mountainous nature. Flash floods in such small, steep watersheds can further be conceptualized to occur as debris flows, hyperconcentrated flows or clear water flows [Costa, 1988, Hyndman and Hyndman, 2008, Jakob et al., 2004]. Costa [1988] differentiates: (i) clear water floods as newtonian, turbulent fluids with non-uniform concentration profiles and sediment concentrations of less than about 20% by volume and shear strengths less than 10 N/m^2; (ii) hyperconcentrated flows as having sediment concentrations ranging from 20 to 47% by volume and shear strengths lower than about 40 N/m^2; and (iii) debris flows as being non-Newtonian visco-plastic or dilatant fluids with laminar flow and uniform concentration profiles, with sediment concentrations ranging from 47 to 77% by volume and shear strengths greater than about 40 N/m^2. Debris flow dominated areas can be subject to hyperconcentrated flows as well as clear water floods [Larsen et al., 2001, Lavigne and Suwa, 2004, Santo et al., 2015], depending on the hydroclimatic conditions and the availability of sediments [Jakob and Weatherly, 2005], and occurrence of all types in the same watersheds has been reported [Larsen et al., 2001, Santo et al., 2015]. Therefore, the areas exposed to clear water floods and debris flows were combined. This provides a conservative delineation of the areas considered to be exposed to flooding.

Exposure areas were obtained from an analysis of the susceptibility to flooding. Areas that potentially can be affected by clear water floods and debris flows were determined using simplified methods that provide a mask where the analysis of exposed elements was carried out. The probability of occurrence and magnitude are not considered in the analysis, since the scope of the simplified regional assessment is limited to assessing the susceptibility of the watersheds to flooding. Areas prone to debris flows were previously identified by Rogelis and Werner [2013] through application of the Modified Single Flow Direction model.

In order to delineate the areas prone to clear water floods, or floodplains, two geomorphic-based methods were tested using a digital elevation model with a pixel size of 5 metres as an input, which was obtained from contours. Floodplains are areas near stream channels shaped by the accumulated effects of floods of varying magnitudes and their associated geomorphological processes. These areas are also referred to as valley bottoms and riparian areas or buffers [Nardi et al., 2006].

The first approach is the multi-resolution valley bottom flatness (MRVBF) algorithm [Gallant and Dowling, 2003]. The MRVBF algorithm identifies valley bottoms using a slope classification constrained on convergent area. The classification algorithm is applied at multiple scales by progressive generalisation of the digital elevation model, combined with progressive reduction of the slope class threshold. The results at different scales are then combined into a single index. The MRVBF index utilises the flatness and lowness characteristics of valley bottoms. Flatness is measured by the inverse of slope, and lowness is measured by ranking the elevation with respect to the surrounding area. The two measures, both scaled to the range 0 to 1, are combined by multiplication and could be interpreted as membership functions of fuzzy sets. While the MRVBF is a continuous measure, it naturally divides into classes corresponding to the different resolutions and slope thresholds [Gallant and Dowling, 2003].

In the second method considered, threshold buffers are used to delineate floodplains as areas contiguous to the streams based on height above the stream level. Cells in the digital elevation model adjacent to the streams that meet height thresholds are included in the buffers [Cimmery, 2010]. Thresholds for the height of 1, 2, 3, 4, 5, 7 and 10 metres were tested.

In order to evaluate the results of the MRVBF index and the threshold buffers, flood maps for the study area were used. These are available for only 9 of the 106 watersheds, and were developed in previous studies through hydraulic modelling for return periods up to 100 years. The delineation of the flooded area for a return period of 100 years was used in the nine watersheds to identify the suitability of the floodplain delineation methods to be used in the whole study area. With respect to areas prone to debris flows, these were validated with existing records in the study area by Rogelis and Werner [2013].

3.3.2.2 Choice of indicators and principal component analysis for vulnerability assessment

In this study vulnerability in the areas identified as being exposed is assessed through the use of indicators. The complexity of vulnerability requires a transformation of available data to a set of important indicators that facilitate an estimation of vulnerability [Birkmann, 2006]. To this end, principal component analysis was applied to variables describing vulnerability in the study area in order to create composite indicators [Cutter et al., 2003]. The variables were chosen by taking into account their usefulness according to the literature, and were calculated using the exposure areas as a mask.

Table 3.2 shows the variables chosen to explain vulnerability in the study area. These are grouped in *socio-economic fragility, lack of resilience and coping capacity* and *physical exposure*. The variables are classified according to their social level (individual, household, community and institutional), hazard dependence and influence on vulnerability (increase or decrease). The third column specifies the spatial aggregation level of the available data. The three spatial levels considered are urban block, watershed and locality, where the locality corresponds to the 20 administrative units of the city. The data used to construct the indicators was obtained from the census and reports published by the municipality. For each variable the values were normalised between the minimum and the maximum found in the study area. In the case of variables that contribute to decreasing vulnerability a transformation was applied so a high variable value represents high vulnerability for all variables.

TABLE 3.2: Variables used to construct vulnerability indicators

Social levels	Variable	Effect	spatial level	Description
			Socio-economic Fragility	
Individual	Age		Urban block	Percentage <10 plus percentage >65
	Disability		Urban block	% of population having any sort of disability
	Unemployment		Locality	Unemployement rate
	Income		Locality	Unsatisfied basic needs index - UBN, % of homeless, % of poor population
	Life expectancy		Locality	Life expectancy
	Household size		Locality	Average number of persons per household
	Woman-headed households		Locality	Percentage of families headed by women
Community	Illegal settlements		Urban block	Percentage of illegal settlements
	% of population of strata 1 and 2		Urban block	The socio-economic stratification system of Bogotá classifies the population into strata with similar economic characteristics on a scale from 1 to 6, with 1 as the lowest income area and 6 as the highest. Strata 1 and 2 corresponds to the socio-economic classification with the lowest income.
	Life conditions		Locality	Life conditions index
	Human development index		Locality	Human development index
	Demographic pressure		Locality	Population growth rate
	Child mortality		Locality	Chid mortality rate
Institutional	---	---	---	---
			Lack of Resilience and coping capacity	
Individual	Level of Education		Locality	% of population with education level superior to high school
	Illiteracy		Locality	Illiteracy rate
	Access to information		Locality	% of homes with internet access
Household	---	---	---	---
Community	Risk perception		Watershed	Boolean indicator. A value of 1 was assigned to watersheds where floods have occurred previously and 0 if they have not.
	Robberies		Locality	Robberies per 10000 inhabitants. The robberies that occur in the locality of the watershed were used as a proxy for trust, confidence and the level at which a proper post-disaster environment could be expected, since a high probability of crime can affect the evacuation procedures and the process to recover.
	Participation		Locality	Percentage of eligible voters that voted in the most recent communal elections.
Institutional	Infrastructure/ accessibility		Locality	% of roads in good condition
	Early warning		Watershed	Boolean indicator. Existence of flood early warning systems in the watershed. Watersheds where flood early warning systems are operational were assigned a value of 1 and 0 if they do not exist.
	Hospital beds		Locality	Hospital beds per 10000 inhabitants
	Health care HR		Locality	Health care human resources per 10000 inhabitants
	Rescue personnel		Locality	Rescue personnel per 10000 inhabitants.
			Physical exposure	
Individual	Population exposed		Urban block	Number of people in flood-prone areas
	Density of population		Urban block	People per km^2 in flood-prone areas.
Household	Residential units		Urban block	Number of houses in flood-prone area
Community	Commercial and industrial units		Urban block	Number of commercial and industrial establishments in flood-prone area.
	Community infrastructure		Urban block	Number of community, social, cultural, health care infrastructure exposed
Institutional	---	---	---	---

Legend:	Variable		Effect	
		Hazard dependent		Increases vulnerability
		Hazard independent		Reduces vulnerability

In order to construct the composite indicators related to *socio-economic fragility* and *physical exposure*, principal component analysis (PCA) was applied on the corresponding variables shown in Table 3.2. PCA reduces the dimensionality of a data set consisting of a large number of interrelated variables, while retaining as much as possible of the variation present in the data set. This is achieved by transforming to a new set of variables, the principal components (PCs), which are uncorrelated [Jolliffe, 2002]. The number of components to be retained from the PCA was chosen by considering four criteria: the Scree test acceleration factor, optimal coordinates [Cattell, 1966], the Kaiser's eigenvalue-greater-than-one rule [Kaiser, 1960] and parallel analysis [Horn, 1965]. Since the number of components may vary among these criteria, the interpretability was also taken into account when selecting the components to be used in further analysis, with each PC being considered an intermediate indicator. Subsequently a varimax rotation [Kaiser, 1958] was applied to minimise the number of individual indicators that have a high loading on the same principal component, thus obtaining a simpler structure with a clear pattern of loadings [Nardo et al., 2008]. The intermediate indicators (PCs) were aggregated using a weight equal to the proportion of the explained variance in the data set [Nardo et al., 2008] to provide an overall indicator for *socio-economic fragility* and for *physical exposure*.

PCA has the disadvantage that correlations do not necessarily represent the real influence of the individual indicators and variables on the phenomenon being measured [Nardo et al., 2008]. This can be addressed by combining PCA weights with an equal weighing scheme for those variables where PCA does not lead to interpretable results [Esty et al., 2006]. In the construction of the *lack of resilience and coping capacity* indicator, this issue led to a separation of variables in four groups:

- Robberies and participation: These were treated separately from the rest of the variables to maintain interpretability as a measure of cohesiveness of the community. *Cohesiveness of the community* was identified as a factor that influences the resilience since the degradation of social networks limits the social organisation for emergency response [Ruiz-Pérez and Gelabert Grimalt, 2012]. Since there are only two variables to measure this aspect of resilience, PCA was not applied, and the average of the variables was used instead.

- Risk perception and early warning: Risk perception depends on the occurrence of previous floods, thus it depends on hazard exclusively. The existence of early warning is manly an institutional and organizational issue. Therefore, an interpretation of correlation of these variables with other variables in the group of *lack of resilience and coping capacity* is not possible. These variables were considered separated intermediate indicators. *Risk perception* and *early warning* decrease the lack of coping capacity [Molinari et al., 2013], and therefore an equal negative weight was assigned to these indicators summing up to -0.2. This value was chosen so that their combined influence is less than the individual weight of the other four indicators. The sensitivity of this subjective choice was tested. The effectiveness of flood early warning is closely related to the level of preparedness as well as the available time for implementation of appropriate actions [Molinari et al., 2013]. Due to the rapid hydrologic response and configuration of the watersheds in the study area, flood early warning actions are targeted at reducing exposure and vulnerability and not at hazard reduction.

- Rescue personnel: this variable was initially used in the PCA with all *lack of resilience and coping capacity* variables. However, it was found to increase with *lack of resilience and coping capacity*. This implied that the statistical behaviour of the variable did not represent its the real influence on vulnerability. It was therefore treated independently.

- *Level of education, illiteracy, access to information, infrastructure/accessibility, hospital beds* and *health care HR*: PCA was applied to these variables, since they exhibit high correlation and are interpretable in terms of their influence on vulnerability.

To combine all the *lack of resilience and coping capacity* intermediate indicators into a composite indicator, weights summing up to 1 were assigned (see Section 4.3 for an explanation of the resulting intermediate indicators).

The indicators corresponding to *socio-economic fragility, lack of resilience and coping capacity* and *physical exposure* were combined, assigning equal weight to the three components, to obtain an overall *vulnerability indicator*. The watersheds were subsequently categorised as being low, medium or high vulnerability based on the value of the *vulnerability indicator* and using equal intervals. This method of categorisation was chosen to avoid dependence on the distribution of the data, so monitoring of evolution in time of vulnerability can be carried out applying the same criteria.

3.3.2.3 Sensitivity of the vulnerability indicator

The influence of all subjective choices applied in the construction of the indicators was analysed. This included both choices made in the application of PCA, and for the weighting scheme adopted for the factors contributing to resilience and total vulnerability.

1. For the application of PCA, sensitivity to the following choices was explored:

   (a) Four alternatives for the number of components to be retained were assessed as explained in Section 3.2.2.

   (b) Five different methods in addition to the varimax rotation were considered: Unrotated solution; quatimax rotation [Carroll, 1953, Neuhaus, 1954]; promax rotation [Hendrickson and White, 1964]; oblimin [Carroll, 1957]; simplimax [Kiers, 1994]; and cluster [Harris and Kaiser, 1964].

2. For the weighting scheme, sensitivity to the following choices was explored:

(a) The weights used in the four groups of variables that describe *lack of resilience and coping capacity* were varied by ± 10%.

(b) The weights used to combine the three indicators that result in the final vulnerability composite indicator were varied by ± 10%.

All possible combinations were assessed and the results in terms of the resulting vulnerability category (high, medium and low) were compared in order to identify substantial differences as a result of the choices of subjective options.

3.3.2.4 Categories of recorded damage in the study area

A database of historical flood events compiled by the municipality was used to classify the watersheds in categories, depending on damages recorded in past flood events. For each of these events the database includes: date, location, injured people, fatalities, evacuated people, number of affected houses and an indication of whether the flow depth was higher than 0.5 m or not. Unfortunately, no information on economic losses is available and as the database only covers the period from 2000 to 2012 it is not possible to carry out a frequency analysis. Complete records were only available for 14 watersheds. The event with the highest impact for each watershed was chosen from the records. Subsequently, the 14 watersheds were ordered according to their highest impact event. The criteria to sort the records and to sort the watersheds according to impact from highest to lowest were the following (in order of importance):

1. Fatalities

2. Injured people

3. Evacuated people

4. Number of affected houses

Watersheds with similar or equal impact were grouped, resulting in 11 groups. The groups were again sorted according to damage. A score from 0 to 10 was assigned, where a score of 0 implies that no flood damage has been recorded in the watershed for a flood event, despite the occurrence of flooding, while a score of 10 corresponds to watersheds where fatalities or serious injuries have occurred (see Table 3.3). The 11 groups were further classified into three categories according to the emergency management organization that was needed for the response: (i) low: the response was coordinated locally; (ii) medium: centralized coordination is needed for response with deployment of resources of mainly the emergency management agency; (iii) high: centralized coordination is needed with an interistitutional response. This classification was made under the assumption that the more resources

TABLE 3.3: Categories of recorded damage

Category	Score	Description
Low	0	No recorded damage in the watershed.
Low	1	Events that affect 1 house without causing injuries or fatalities and without the need of evacuation.
Low	2	Events that affect 1 house without causing injuries or fatalities and with the need of evacuation.
Low	3	Events that affect up to 5 houses without causing injuries or fatalities, flood depth less than 0.5 m with evacuation of families.
Medium	4	Events that affect up to 5 houses without causing injuries or fatalities, flood depth higher than 0.5 m with evacuation of families.
Medium	5	Events that affect up to 10 houses without causing injuries or fatalities with evacuation of families.
Medium	6	Events that affect 10-20 houses without causing injuries or fatalities with evacuation of families, flood depth less than 0.5 m.
High	7	Events that affect 10-20 houses without causing injuries or fatalities with evacuation of families, flood depth higher than 0.5 m.
High	8	Events that affect 20-50 houses without causing injuries or fatalities with evacuation of families and possibility of structural damage in the houses.
High	9	Events that affect more than 50 houses without causing injuries or fatalities with evacuation of families and possibility of structural damage in the houses.
High	10	Events that cause fatalities or injuries.

are needed for response the more severe the impacts are, allowing in this way a comparison with three levels of priority classification.

3.3.2.5 Prioritization of watersheds

Due to the regional character and scope of the method applied in this study, a qualitative proxy for risk was used to prioritise the watersheds in the study area. A high priority indicates watersheds where flood events will result in more severe consequences. However, the concept of probability of

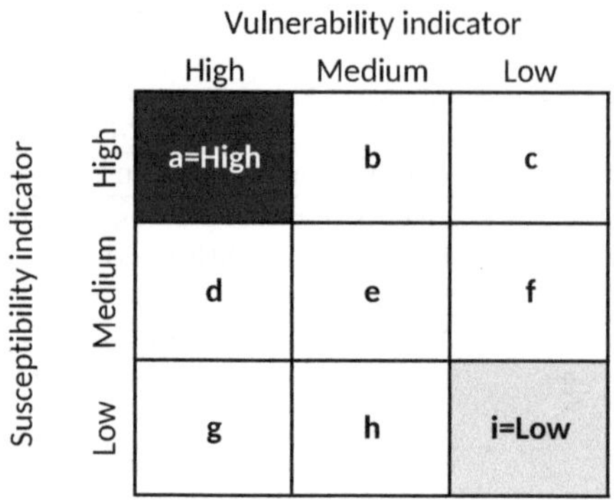

FIGURE 3.2: Initial matrix of priority

occurrence of these is not involved in the analysis, since the analysis of flood hazard is limited to susceptibility.

In order to combine the vulnerability and susceptibility to derive a level of risk, a classification matrix was used. This is shown in Figure 3.2. The columns indicate the classification of the vulnerability indicator and the rows the classification of the susceptibility indicator. Only two priority outcomes are well defined, these are the high and low degrees assigned to the corners of the matrix corresponding to high susceptibility and high vulnerability and low susceptibility and low vulnerability (cells a and i), since they correspond to the extreme conditions in the analysis. The priority outcomes in cells from b to h were considered unknown and to potentially correspond to any category (low, medium or high priority). To define the category for these cells, the priority using all possible matrices (all possible combinations of categories of cells b to c) was assessed for the watersheds for which flood records are available. Once, these watersheds were prioritised, a contingency table is constructed comparing the priority with the damage category (from Table 3.3) from which the "proportion correct" is obtained. The classification matrix that results in the highest proportion correct (best fit) was used for the prioritisation of the whole study area.

3.4 Results

3.4.1 Exposure Areas

Figure 3.3 shows the results of the methods applied to identify areas susceptible to flooding through clear water floods or debris flows. Figure 3.3-a shows the debris flow propagation extent derived for the watersheds of the Tunjuelo basin and the Eastern Hills by Rogelis and Werner [2013]. Since the method does not take into account the volume that can be deposited on the fan, this shows the

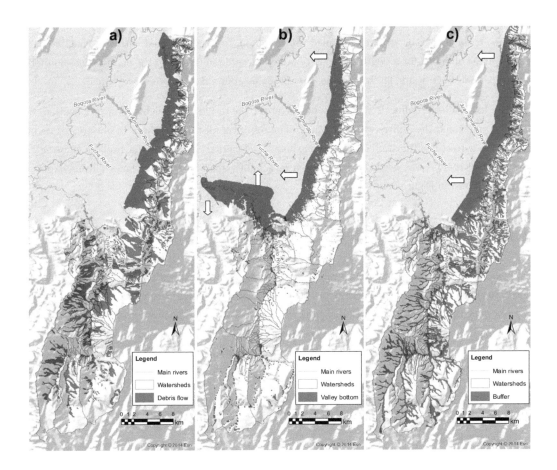

FIGURE 3.3: Clear water flood and debris flow susceptibility areas. Areas in dark grey in each map represent; a) debris flow extent [Rogelis and Werner, 2013]; b) Valley bottoms identified using the the MRVBF index; c) Buffers. In the case of maps b and c, the flood prone areas extend in the direction of the arrows over the flat area.

maximum potential distance that the debris flow could reach according to the morphology of the area, which is in general flat to the west of the Eastern Hills watersheds. A different behaviour can be observed in the watersheds located in the Tunjuelo river basin where the marked topography and valley configuration restricts the propagation areas.

Figure 3.3-b shows the results obtained from the MRVBF index. The comparison of the index with the available flood maps in the study area shows that values of the MRVBF higher than 3 can be considered areas corresponding to valley bottoms. In areas of marked topography the index identifies areas adjacent to the creeks in most cases and the larger scale valley bottoms. However, in flat areas the index unavoidably takes high values and cannot be used to identify flood prone areas.

Figure 3.3-c shows the result obtained from the use of buffer thresholds. The buffers that were obtained by applying the criteria explained in Section 3.2.1, were compared with the available flood maps. Areas obtained for a depth criterion of 3 meters were the closest to the flood delineation for a return period of 100 years, and this value was chosen as appropriate for the study area.

In order to obtain the delineation of the exposure areas, the results of the debris flow propagation; the MRVBF index and the buffers were combined. The results of all three methods in flat areas does not allow for a correct identification of flood prone areas, and a criteria based on the available information and previous studies was needed to estimate a reasonable area of exposure. The resulting exposure areas are shown in Figure 3.4.

3.4.2   Socio-economic fragility indicators

The results of the principal component analysis applying a varimax rotation are shown in Table 3.4. Two principal components were retained as this allowed a clear interpretation to be made for each of the components. The variables included in the first principal component are related to *lack of well-being* (P_{LofW}), while in the second these are related to the *demography* (P_{demog}). The two principal components account for 79 percent of the variance in the data with the first component explaining 80% of the variance (PVE) and the second 20%.

Using the factor loadings (correlation coefficients between the PCs and the variables) obtained from the analysis (see Table 3.4) and scaling them to unity, the coefficients of each indicator are shown in the following equations:

$$P_{LofW} = 0.10Whh + 0.10UE + 0.10PUBNI +$$
$$0.09Ho + 0.11P + 0.10Pho + 0.09M +$$
$$0.10LE + 0.08QLI + 0.10HDI + 0.04G$$

$$(3.1)$$

$$P_{demog} = 0.32Age + 0.20D + 0.29PE12 + 0.19IS$$

$$(3.2)$$

The impacts of the indicators imply that the higher the *lack of well-being* the higher the *socio-economic fragility*, and equally the higher the *demography indicator* the higher the *socio-economic fragility*. Using the percentage of variability explained (PVE) by each component, the composite indicator for *socio-economic fragility* (P_{soc-ec}) is found as:

$$P_{soc-ec} = 0.8P_{LofW} + 0.2P_{demog}$$

$$(3.3)$$

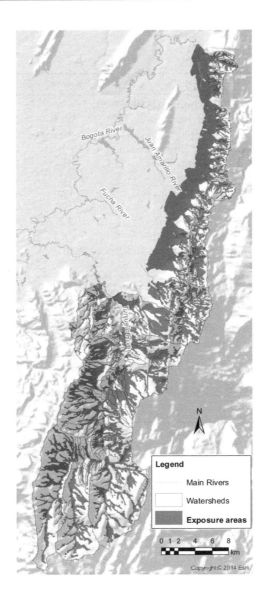

FIGURE 3.4: Exposure areas

3.4.3 Lack of Resilience and coping capacity indicators

The loadings of the indicators representing *lack of resilience and coping capacity* obtained from the PCA are shown in Table 3.5. Two principal components were used; the first correlated with variables related to the *lack of education* (P_{LEdu}) and the second with variables related to *lack of preparedness*

TABLE 3.4: Results of the principal component analysis for socio-economic fragility indicators.

Variable	Symbol	Loadings
Lack of Well-being (PVE=0.8)		
Women-headed households	Whh	0.94
Unemployment	UE	0.97
Poor-Unsatisfied Basic Needs Index	PUBNI	0.98
% Homeless	Ho	0.92
% Poor	P	0.99
Persons per home	Pho	0.94
Mortality	M	0.91
Life Expectancy	LE	0.94
Quality life index	QLI	0.86
Human Development Index	HDI	0.97
Population Growth Rate	G	0.57
Demography (PVE=0.2)		
% of Children and Elderly	Age	0.84
% Disabled	D	0.67
% Population estrata 1 and 2	PE12	0.81
% Illegal settlements	IS	0.64

TABLE 3.5: Results of the principal component analysis resilience indicators

Variable	Symbol	Loadings
Lack of Education (PVE=0.53)		
Level of Education	LEd	0.94
Illiteracy	I	0.96
Access to information	AI	0.93
Lack of Prep. and Resp. Capacity (PVE=0.47)		
Infrastructure/accessibiliy	IA	0.80
Hospital beads	Hb	0.97
Health Care HR	HRh	0.92

and *response capacity* ($P_{LPrRCap}$). These account for 97 percent of the variance in the data with the first component explaining 53% of the variance (PVE) and the second 47%.

Using the factor loadings obtained from the analysis and scaling them to unity, the coefficients of each indicator are shown in the following equations:

$$P_{LEdu} = 0.33LEd + 0.32I + 0.35AI \qquad (3.4)$$

$$P_{LPrRCap} = 0.26IA + 0.39Hb + 0.35HRh \tag{3.5}$$

In an initial analysis, the variable *rescue personnel* was included in the principal component analysis. Results showed a high negative correlation of this variable with *lack of education, illiteracy* and *access to information*. This may be due to more institutional effort being allocated to deprived areas that are more often affected by emergency events in order to strengthen the response capacity of the community. Also civil protection groups rely strongly on voluntary work that seems to be more likely in areas with lower education levels.

Since the consideration of *rescue personnel* changes the interpretation of the principal component that groups the *lack of education* and *access to information* indicator, it was decided to exclude it from the PCA and to consider this variable as an independent indicator (*Lack of Rescue Capacity*).

In the analysis of *robberies* and *participation* as variables describing *cohesiveness of the community*, it was found that the increase in crime is correlated with the lack of *participation*, describing the distrust of the community both of neighbours and of institutions. The corresponding composite indicator was calculated as the average of *robberies* and lack of *participation*.

The equation of *Lack of Resilience and coping capacity* is shown in equation 3.6. Equal weight was assigned to the indicators reflecting *Lack of Education, Lack of Preparedness and Response Capacity, Lack of Rescue Capacity* (P_{LRc}) and *Cohesiveness of the Community* (P_{CC}); and a weight of -0.1 to *Risk Perception* (P_{RP}) and *Early Warning* (P_{FEW}).

$$P_{LRes} = 0.25P_{LEdu} + 0.25P_{LPrRCap} + 0.25P_{LRc}$$
$$0.25P_{CC} - 0.1P_{RP} - 0.1P_{FEW} \tag{3.6}$$

The indicator of *lack of resilience and coping capacity* was obtained it was rescaled between 0 and 1.

3.4.4   Physical exposure indicators

The principal component analysis of the variables selected for *physical exposure* shows that these can be grouped into two principal components that explain 82% of the variability (exposed infrastructure - P_{Ei} and exposed population - P_{Ep}). The results of the analysis are shown in Table 3.6.

Using the factor loadings obtained from the analysis and scaling them to unity, the coefficients of each composite indicator are shown in the following equations:

$$P_{Ei} = 0.32Ncb + 0.37Niu + 0.32Ncu \tag{3.7}$$

TABLE 3.6: Results of the principal component analysis physical susceptibility indicators

Variable	Symbol	Loadings
Exposed infrastructure (PVE=0.52)		
Number of civic buildings	Ncb	0.86
Number of industrial units	Niu	0.96
Number of comercial units	Ncu	0.85
Exposed population (PVE=0.48)		
Number of residential units	Nru	0.91
Population exposed	Pe	0.85
Density of population	Dp	0.78

$$P_{Ep} = 0.38Nru + 0.33Pe + 0.28Dp \tag{3.8}$$

Using the percentage of variability explained (PVE) by each indicator, the composite indicator of *physical susceptibility* is found to be:

$$P_{ps} = 0.52P_{Ei} + 0.48P_{Ep} \tag{3.9}$$

3.4.5 Vulnerability indicator

The resulting *vulnerability indicator* was obtained through the equal-weighted average of the indicators for *socio-economic fragility*, *lack of resilience and coping capacity*, and *physical exposure*. Categories of low, medium and high vulnerability for each watershed were subsequently derived based on equal bins of the indicator value. The spatial distribution is shown in Figure 3.5, as well as the spatial distribution of the three constituent indicators.

Conditions of *lack of well-being* are shown to be concentrated in the south of the study area. The *demographic conditions* are more variable, showing low values (or better conditions) in the watersheds in the South, where the land use is rural. Low values also occur in the North, where the degree of urbanization is low due the more formal urbanization processes (see Figure 3.5-a). The spatial distribution of the indicator of *lack of resilience and coping capacity* (Figure 3.5-b) shows that the highest values are concentrated in the south-west of the study area where the education levels are lower and the road and health infrastructure poorer. The same spatial trend is exhibited by the *lack of preparedness and response capacity*. The south of the study area corresponds mainly to rural use, thus the *physical exposure* indicator shows low values (see Figure 3.5-a). The highest values are concentrated in the centre of the area where the density of population is high and the economic activities are located.

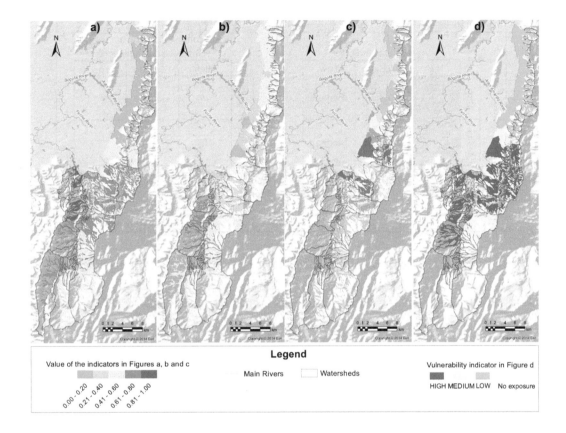

FIGURE 3.5: a) Spatial distribution of the Socio-economic indicator; b) Spatial distribution of the resilience indicator; c) spatial distribution of the physical exposure indicator; d) Spatial distribution of the total vulnerability indicator

The spatial distribution of the overall indicator and the derived categories show that the high vulnerability watersheds are located in the centre of the study area and in the west.

3.4.6   Prioritization of watersheds according to the qualitative risk indicator and comparison with damage records

The "proportion correct" of all possible matrices according to Section 3.2.5 (see Figure 3.2) resulted in the optimum matrix shown in Figure 3.6-a, the corresponding contingency matrix is shown in Figure 3.6-b with a "proportion correct" of 0.85.

The prioritisation level obtained from the application of the combination matrix to the total *vulnerability indicator* and the *susceptibility indicator* for each watershed is shown in Figure 3.7-a.   The

a) Vunerability-susceptibility combination matrix

Susceptibility indicator \ Vulnerability indicator	High	Medium	Low
High	High	Medium	Medium
Medium	High	Medium	Medium
Low	Low	Low	Low

b) Contingency matrix

i. 3x3 contingency matrix

Predicted priority \ Observed damage score	High	Medium	Low
High	7	0	0
Medium	0	3	2
Low	0	0	1

Proportion correct = 0.85

FIGURE 3.6: a) Vulnerability-Susceptibility combination matrix. b) Contingency matrix.

results were assigned to the watersheds delineated up to the discharge into the Tunjuelo River or into the storm water system, in order to facilitate the visualisation. The damage categorisation of the study area using the database with historical records according to Table 3.3 is shown in Figure 3.7-b with range categories classified as high, medium and low. This shows that the most significant damages, corresponding to the highest scores for the impact of flood events, are concentrated in the central zone of the study area. The comparison between Figure 3.7-a and Figure 3.7-b shows that the indicators identify a similar spatial distribution of priority levels in the central zone of the study area that is consistent with the distribution of recorded damage. This is reflected in the "proportion correct" of 0.85.

3.4.7 Sensitivity analysis of the vulnerability indicator

Figure 3.8 shows the box plots of the values of the vulnerability indicator obtained from the sensitivity analysis in application of PCA as well as the weighting scheme as explained in Section 3.2.3. The values of the vulnerability indicator obtained from the proposed method were also plotted for reference. The most influential input factors correspond to the weights used both in the construction of the lack of resilience indicator and in the construction of the total vulnerability indicator. The thick vertical bars for each watershed show the interquartile range of the total vulnerability indicator, with the thin bars showing the range (min-max). While the range of the indicator for some watersheds is substantial, the sensitivity of the watersheds being classified differently in terms of low, medium or high vulnerability was evaluated through the number of watersheds for which the interquartile range intersects with the classification threshold. For seven watersheds classified as of medium vulnerability the interquartile range crosses the upper limits of classification of medium vulnerability, while for four

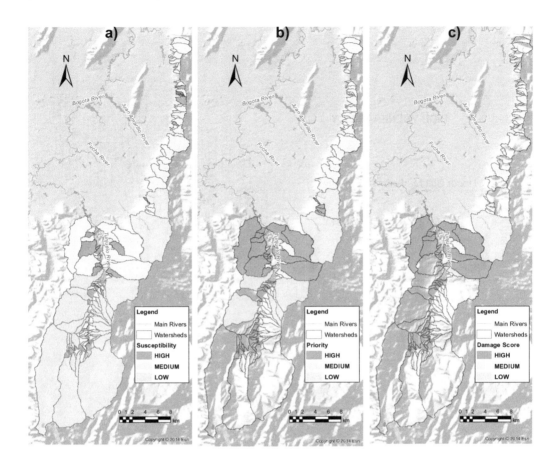

FIGURE 3.7: a) Susceptibility classification of the study area. b) Prioritisation according to the qualitative risk indicator. c) Damage categorization

watersheds classified as of high vulnerability the range crosses that same threshold. For the lower threshold, only two watersheds classified as being of low vulnerability are sensitive to crossing into the class of medium vulnerability.

3.5 Discussion

3.5.1 Exposure areas

Existing flood hazard maps developed using hydraulic models that were available for a limited set of the watersheds in the study area were used to assess the suitability of the proposed simplified methods

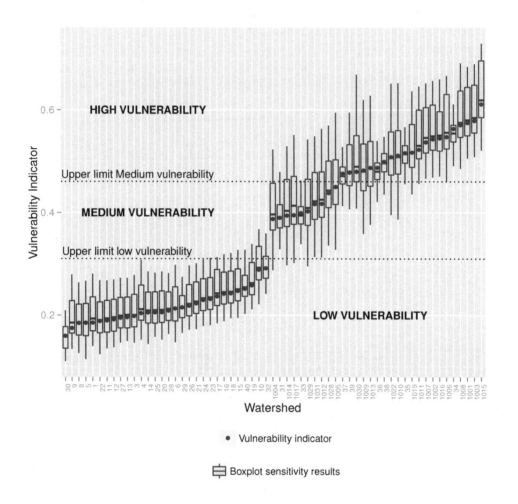

FIGURE 3.8: Sensitivity analysis of the vulnerability indicator. Note: The numbering of the watersheds in the Eastern Hills goes from 1 to 40 and in the Tunjuelo River Basin from 1000 to 1066.

to identify flood prone areas and extend the flood exposure information over the entire study area. The areas exposed to debris flows obtained through the Modified Single-Flow Direction Propagation algorithm show a good representation of the recorded events [Rogelis and Werner, 2013]. However, in the eastern hills, where the streams flow towards a flat area, the results of the algorithms tend to overestimate the propagation areas since in these algorithms the flood extent is dominated purely by the morphology and the flood volume is not considered, which means there is no limitation to the flood extent (see Figure 3.3).

Each of the methods applied for flood plain delineation has strengths and weaknesses, while the combination of the results from these methods provides a consistent and conservative estimate of the exposure areas. The MRVBF index allows the identification of valley bottoms at several scales. In the mountainous areas, zones contiguous to the streams are identified, and in areas of marked topography the results are satisfactory, allowing a determination of a threshold of the index to define flood prone areas. In the case of the buffers (see Figure 3.3-c), a depth of 3 meters seems adequate to represent the general behaviour of the streams. The combination of the methods allowed the estimation of exposure areas based on the morphology (low and flat areas), elevation difference with the stream level (less than 3 meters) and capacity to propagate debris flows.

3.5.2   Representativeness and relative importance of indicators

The principal component analysis of the variables used to explain *socio-economic fragility* showed that the 16 variables that were chosen for the analysis could be grouped into two principal components strongly associated with the *demography* and the *lack of well-being* in the area. The latter was found to explain most of the variance in the data (80 % as shown in Table 3.4).

The *demography* intermediate indicator describes the dependent population and the origin of the population. Dependent population (children, elderly and disabled) has been also identified by other authors as an important descriptor of vulnerability [Cutter et al., 2003, Fekete, 2009], associated to the limited capacity of this population to evacuate [Koks et al., 2015] and recover [Rygel et al., 2006]. The origin of the population (illegal settlements and % of population in strata 1 and 2) shows the proportion of population resulting mainly from forced migration due to both violence and poverty [Beltrán, 2008].

The *lack of well-being* indicator is composed of 14 strongly correlated variables that are commonly used to measure livelihood conditions. Poverty does not necessarily mean vulnerability, though the lack of economic resources is associated with the quality of construction of the houses, health and education, which are factors that influence the capability to face an adverse event [Rygel et al., 2006]. The variable *"women-headed households"* is correlated with the principal component related to *lack of well-being* as identified by Barrenechea et al. [2000]. Even if this condition of the families is not necessarily a criteria related to poverty, women-headed households with children are related to vulnerability conditions. The woman in charge of the family is responsible for the economic, affective and psychological well-being of other persons, specially her children and elderly, in addition to domestic tasks and the family income. This condition suggest more assistance during emergency and recovery [Barrenechea et al., 2000].

In the case of the *lack of resilience and coping capacity* indicators, the PCA resulted in the intermediate indicators *lack of education* and *lack of preparedness and response capacity*. The former captures limitations in knowledge about hazards in individuals [Müller et al., 2011] and the latter is linked to the institutional capacity for response. Risk perception and early warning are boolean indicators. Since risk perception is based on the occurrence or non-occurrence of floods, aspects such as specific knowledge of the population about their exposure are not included. In the case of flood early warning, the effectiveness of the systems is not considered. These are aspects that can be taken into account for future research and that can help to improve the *lack of resilience and coping capacity* indicators.

Regarding the *physical exposure*, the method that was applied does not involve hazard intensity explicitly and different levels of physical fragility are not considered due to limitations in the available data. The indicators used to express physical exposure imply that the more elements exposed the more damage, neglecting the variability in the degree of damage that the exposed elements may have. Other regional indicator-based approaches have used physical characteristics of the exposed structures to differentiate levels of damage according to structure type [Kappes et al., 2012] and economic values of the exposed elements [Liu and Lei, 2003]. This is a potential area of improvement of the indicator, since the degree of damage depends on the type and intensity of the hazard and the characteristics of the exposed element. However, the development of indicators of physical characteristics and economic values is highly data demanding, therefore future applications could be aimed at efficiently use existing information and apply innovative data collection methods at regional level for the improvement of the physical indicator.

3.5.3 Sensitivity of the vulnerability indicator

The interquartile ranges cross the thresholds between categories of low, medium and high vulnerability only in the case of 13 watersheds (see Figure 3.8). This means that only these 13 watersheds are sensitive to the criteria selected for the analysis. In 11 of these, the category changes between medium vulnerability and high vulnerability and in the remaining two the change is from low to medium vulnerability. Watersheds with values of the *vulnerability indicator* out of the intermediate ranges of the thresholds are robust to the change in the modelling criteria. Clearly, these results are dependent on the number of categories. While introducing more categories may provide more information to differenciate watersheds, the identification of category of the watersheds may become more difficult due to the sensitivity to the results. Therefore, in order to preserve identifiability of the vulnerability category of the watersheds more than three categories could not be used. Indicator-based regional studies that classify vulnerability in 3 categories, have shown to provide useful information for flood risk management [Kappes et al., 2012, Liu et al., 2015, Luino et al., 2012].

The impact on the proportion correct of a shift of category for the 13 watersheds mentioned above can only be assesssed for the 2 watersheds where flood records are available. This does not result in changes in the contingency matrix shown in Figure 3.6-b. With respect to the assigning the priority to the watersheds, only 7 (7% of the total) of the 13 watersheds that showed sensitivity to a shift of the vulnerability categories were found to be sensitive to a change in priority (high/medium), which reflects the robustness of the analysis using the considered categories.

3.5.4   Usefulness of the prioritization indicator

The resulting vulnerability-susceptibility combination matrix shown in Figure 3.6-a, shows that in the study area high priorities are determined by high vulnerability conditions and medium and high susceptibility. This would suggest that, high vulnerability is a determinant condition of priority, since areas with high vulnerability can only be assigned a low priority if the susceptibility to flash floods/debris flows is low. This also shows that the analysis of the indicators that compose the vulnerability index allows insight to be gained into the drivers of high vulnerability conditions. Figure 3.5 shows that high vulnerability watersheds are the result of:

- High *socio-economic fragility* and high *lack of resilience and coping capacity* (west of the lower and middle basin of the Tunjuelo river; and watershed most to the south of the Eastern Hills).

- High *socio-economic fragility* and high *physical exposure* (east of the middle basin of the Tunjuelo river).

- High *physical exposure* levels (south of the Eastern Hills)

This information is useful for regional allocation of resources for detailed flood risk analysis, with the advantage that the data demand is low in comparison with other indicator-based approaches [Fekete, 2009, Kappes et al., 2012]. Furthermore most weights are determined from a statistical analysis with a low influence of subjective weights, which is an advantage over expert weighting where large variations may occur depending on the expert's perspective [Müller et al., 2011]. However, more detailed flood risk management decision-making cannot be informed by the level of resolution used in this study. Studies where assessments are carried out at the level of house units would be needed for planning of mitigation measures, emergency planning and vulnerability reduction [Kappes et al., 2012]. Although, the proposed procedure could be applied at that more detailed level, this could not be done due to the availability of information. This is a common problem in regional analyses [Kappes et al., 2012] where collecting large amount of data at high resolution is a challenge. Nevertheless, future advances in collection of data could be incorporated in the proposed procedure yielding results at finer resolutions. The challenge not only lies in collecting data of good quality at high resolution that can be transformed into indicators, but also in producing data at the same pace as significant changes in variables that contribute to vulnerability take place in the study area. In this research, vulnerability was assessed statically, however, there is an increasing need for analyses that take into account the dynamic characteristics of vulnerability [Hufschmidt et al., 2005]. Methods such as the one applied in this study can provide a tool to explore these dynamics since it can be adapted to different resolutions according to the data available.

3.6 Conclusions

In this study a method to identify mountainous watersheds with the highest flood risk at the regional level is proposed. Through this, the watersheds to be subjected to more detailed risk studies can be prioritised in order to establish appropriate flood risk management strategies. The method is demonstrated in the steep, mountainous watersheds that surround the city of Bogotá (Colombia), where floods typically occur as flash floods and debris flows. The prioritisation of the watersheds is obtained through the combination of vulnerability with susceptibility to flash floods/debris flows. The combination is carried out through a matrix that relates levels of vulnerability and susceptibility with priority levels.

The analysis shows the interactions between drivers of vulnerability, and how the understanding of these drivers can be used to gain insight in the conditions that determine vulnerability to floods in mountainous watersheds. Vulnerability is expressed in terms of composite indicators; *Socio-economic fragility, lack of resilience and coping capacity* and *physical exposure*. Each of these composite indicators is formed by an underlying set of constituent indicators that reflect the behaviour of highly correlated variables, and that represent characteristics of the exposed elements. The combination of these three component indicators allowed the calculation of a *vulnerability indicator*, from which a classification into high, medium and low vulnerability was obtained for the watersheds of the study area. Tracing back the composite indicators that generate high vulnerability, provided an understanding of the conditions of watersheds that are more critical, allowing these to be targeted for more detailed flood risk studies. In the study area it is shown that those watersheds with high vulnerability are categorised to be of high priority, unless the susceptibility is low, indicating that the vulnerability is the main contributor to risk. Furthermore, the contributing components that determine high vulnerability could be identified spatially in the study area.

The developed methodology can be applied to other areas, although adaptation of the variables considered may be required depending on the setting and the available data. The proposed method is flexible to the availability of data, which is an advantage for assessments in mountainous developing cities and when the evolution in time of variables that contribute to vulnerability is taken into account.

The results also demonstrate the need for a comprehensive documentation of damage records, as well as the potential for improvement of the method. Accordingly, further research should be focused on (i) the use of smaller units of analysis than the watershed scale, which was used in this study; (ii) Improvement of physical exposure indicators incorporating type of structures and economic losses; and (iii) incorporation of more detailed information about risk perception and flood early warning.

Chapter 4

Spatial interpolation for real-time rainfall field estimation in areas with complex topography

This chapter is an edited version of: Rogelis, M. C., and Werner, M. G. F. : Spatial Interpolation for Real-Time Rainfall Field Estimation in Areas with Complex Topography. J. Hydrometeorol., 14, 85–104, doi:10.1175/JHM-D-11-0150.1, 2012

4.1 Introduction

Estimation of runoff for flood assessment and flood early warning requires an appropriate description of the spatial distribution of rainfall. This is particularly the case when distributed rainfall-runoff models are used [Arnaud et al., 2002], where the input scale of the rainfall field should be commensurate with the scale of the model.

Rain gauge networks provide point measurements of rainfall than can be spatially interpolated to estimate the rainfall spatial distribution. Prediction of rainfall over an area with complex topography from rain gauges using geostatistical interpolation offers the advantage over traditional methods such as Thiessen that an estimation of uncertainty is provided.

Selection of the most appropriate interpolation method depends on local rainfall characteristics, the time and spatial scale over which the analysis is carried out, the application for which the data are needed, the rainfall event morphology and the rain gauge network organization [Garcia et al., 2008, Grimes and Pardo-Igúzquiza, 2010]. The nearest neighbor method, the arithmetic mean, spline surface fitting, optimal interpolation, Kriging and interpolation based on empirical orthogonal functions are among the available interpolation techniques. Probabilistic precipitation estimates have been addressed by Clark and Slater [2006] aimed at generating conditional ensemble grids of daily precipitation. The comparison of interpolation techniques by several authors has shown advantages in performance of Kriging procedures over the other methods. Tabios and Salas [1985] compared Kriging with Thiessen, Inverse Distance Weight (IDW), Polynomial trend surfaces and inverse square distance, finding Kriging to be superior. Similar results were found by Buytaert et al. [2006b] when comparing Kriging with Thiessen polygons. Diodato and Ceccarelli [2005] found a better performance of ordinary cokriging in comparison with linear regression and inverse squared distance and Hevesi et al. [1992] concluded that Kriging provided the best results relative to other interpolation methods.

The geostatistical interpolation technique of Kriging groups different methods: Simple Kriging, Ordinary Kriging, Universal Kriging, Kriging with External Drift, Regression Kriging, Intrinsic Kriging etc, depending on the underlying model [Chiles and Delfiner, 1999]. The variogram is a key tool in Kriging methods to represent spatial structure by describing how the spatial continuity changes with distance and direction [Isaaks and Srivastava, 1989].

When interpolation is carried out in real-time, careful determination of the random field structure for each storm event may be too time consuming. For real-time operation with short time steps, reliable values of the model parameters cannot be obtained from the small number of data points commonly available [Lebel and Bastin, 1989]. Climatological variograms constitute an approach with the potential to overcome this difficulty; however, the determination of a climatological variogram requires the rainfall data to be homogeneous, this is, realizations of a unique random field [Lebel and Bastin, 1985]. However, in areas with complex topography or distinct seasonality this may not be the case. Homogeneity can be improved by using zones or seasonal classes. Geostatistical interpolation can then be applied to the data in each class. This classification can be complex, however, particularly when seasonal trends are not evident or when topography is highly complex such as in mountainous areas.

In order to explore the relationship between secondary variables such as elevation and precipitation aimed at improving the interpolated rainfall field, Kriging with external drift was applied. The use of secondary variables has been studied by several authors [Diodato, 2005, Diodato and Ceccarelli, 2005, Guan et al., 2005, Kyriakidis et al., 2001, Marquinez et al., 2003, Vidal and Varas, 1982]. These show the complexity of the relation between precipitation and secondary variables, and address the estimation of precipitation applying regression models and geostatistics. These approaches have been used to estimate rainfall fields for post-event analysis, thus fitting a variogram for the individual events and studying the complex relationship between precipitation and secondary variables for each data set in particular, which may be highly time consuming during real-time operation.

Several authors have pointed out the advantage of using topographic variables as predictors in the interpolation of yearly, monthly and daily rainfall fields, mainly when these are smoothed [Buytaert et al., 2006b, Deraisme et al., 2001, Diodato, 2005, Diodato and Ceccarelli, 2005, Hay et al., 1998, Hevesi et al., 1992, Johansson and Chen, 2003, Ninyerola et al., 2000]. However, one of the topics of discussion is the skill of the procedure to improve the interpolation, in relation to the correlation of the secondary variables with precipitation. Carrera-Hernández and Gaskin [2007] suggest that the interpolation of daily rainfall events is improved by the use of elevation as secondary variable even when the correlation of this variable with precipitation is low. On the other hand Goovaerts [2000] concluded that the benefit of multivariate techniques can become marginal if the correlation between rainfall and elevation (or other environmental descriptors) is too small.

The novelty in this work is the proposal and assessment of a robust procedure for real-time interpolation of point measurements of daily precipitation that can be automated and that is reliable and accurate enough to be used in flood forecasting. The difference in performance that can be expected when using individual variograms and pooled variograms for Ordinary Kriging and Kriging with external drift is examined. This allows to determine if these differences are significant in terms of errors in the rainfall field and in the variation of the precipitation volume over the basin of interest as a critical parameter for runoff estimation. Thus, the potential of simplified interpolation procedures with averaged variograms for automatic interpolation is established, addressing the challenge of defining an adequate spatial structure previous to the occurrence of the precipitation and identifying the contribution of secondary variables in the improvement of the rainfall field. The uncertainty due to choice of interpolation methods on the precipitation volumes is estimated using Gaussian simulation.

In this study, Bogotá (Colombia) was chosen for application of the methodology, since it is located in an area with complex topography and meteorology, influenced by the Intertropical Convergence Zone.

4.2 Methods and Data

4.2.1 Study Area

The study area covers the urban area of Bogotá (Colombia) and the Tunjuelo river basin (Figure 4.1). The bounds of this area define a region were the density of existing rain gauges is high compared to the sparse network in the surrounding areas. Furthermore, the combination of flat and mountainous relief provides the complexity that is needed to test the interpolation methods in a group of watersheds that drain to the Bogotá River for which flood early warning is crucial. The study area covers a surface of $2695 \ km^2$.

Bogotá is the capital of Colombia and is located on a high plateau in the Eastern Andes mountain range of Colombia, with the elevation ranging between 2510 m.a.s.l and 3780 m.a.s.l [Bernal et al., 2007]. The city is surrounded by hills which have limited its growth to the east and south. The

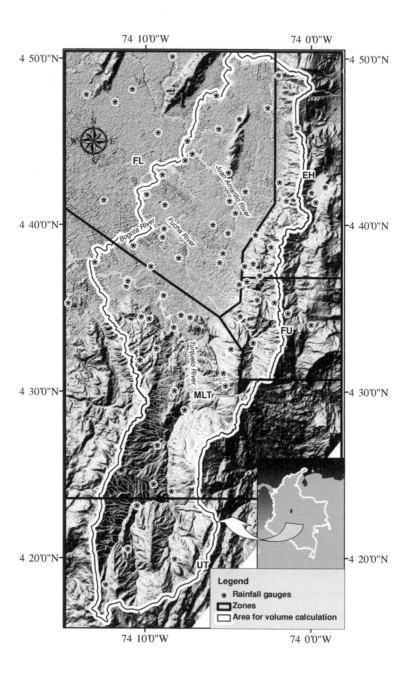

FIGURE 4.1: Study area, location and distribution of rainfall gauges and area for precipitation volume calculation

western city limit is the Bogotá River, which drains a large plain called the Savannah of Bogotá. Because of its location near the Equator, precipitation is influenced by the Intertropical Convergence Zone (ITCZ), defining two rainy seasons (April-May and September-October). Rain may, however, occur throughout the year, with the ITCZ responsible for a continuous moisture supply in the form of rain, clouds and fog, as a result of orographic uplift [Buytaert et al., 2006b].

4.2.2 Precipitation data

Daily data from telemetry and manual stations was collected for the period 2000 to 2009. This choice of time step for the analysis allowed the incorporation of data from all the existing rainfall gauges (automatic, pluviographs and pluviometers).

139 daily data sets with significant precipitation were chosen by identifying daily values above 30 mm. The months with the highest number of daily data sets correspond to April (23 data sets), May (21 data sets) and November (24 data sets), however there are data sets for all the months of the year (January: 8 data sets; February: 15 data sets; March: 16 data sets; June: 9 data sets; July: 3 data sets; August: 3 data sets; September: 6 data sets; October: 5 data sets; December: 6 data sets). The amount of available rainfall gauges ranges from 39 to 85 (depending on the storm), resulting in a density of 1 station per 32 to 69 km^2. Their distribution can be observed in Figure 4.1.

Basic statistic analysis was applied to the 139 daily precipitation datasets to produce a preliminary understanding of the spatial distribution of data. Subsequently, a classification of the datasets was carried out according to the location and extent of the storms.

The assumption of climatological variograms is that all the standardized observations in a given sample are realizations of a unique random field [Lebel and Bastin, 1985]. Due to seasonal and local meteorological conditions, a unique variogram may, however, not be appropriate to describe all the spatial structures of the rainfall fields. Therefore, a classification in terms of location and extent of the rainfall field was carried out. The daily data sets were divided into five classes according to the zone where the location of the maximum daily precipitation was recorded and according to the extent of the precipitation. If the precipitation concentrated mainly in one zone with limited or no rain in the others, the daily dataset was classified as small extent (S). If significant precipitation was recorded in two or more contiguous zones the classification was large extent (L). The zones used for the classification are shown in Figure 4.1. The zones were constructed taking as reference the pluviographic zones of Bogotá [IRH, 1995] but introducing small modifications in such a way that a distinction between flat and mountainous areas was relevant, that zones could be associated with watershed units, and that the zones were as simple as possible. It is expected that this gives similar meteorological characteristics for the daily datasets in such a way that data belonging to the same class of zone and extent can be used to construct the climatological variogram for that zone. The upper part of the Tunjuelo river basin was considered as a separate zone as this area is influenced by the pluviometric regime of the eastern plains of Colombia which implies a difference to the rest of the study area.

4.2.3   Geostatistical interpolation procedure

Five methods for geostatistical interpolation were considered. For the individual storm events ordinary kriging (OK), Universal Kriging (UK) and Kriging with external Drift (KED) were applied. UK is a special case of KED that has been reserved by many authors for the case when only the coordinates are used as predictors. If the deterministic part of variation (drift) is defined externally as a linear function of other explanatory variables, rather than only the coordinates, the term Kriging with External Drift (KED) is preferred. In the case of UK or KED, the predictions are made as with kriging, with the difference that the covariance matrix of residuals is extended with the auxiliary predictors. However, the drift and residuals can also be estimated separately and then summed, then the procedure is named Regression Kriging. Although KED and RK provide equivalent results, they differ in the computational steps used [Hengl, 2009]. In this study the reference will be made to either KED or RK, depending on the procedure applied to solve the kriging system.

KED was carried out including topographic variables smoothed over windows of varying size. As a first step for KED, stepwise regression was undertaken for the chosen topographic variables and with the identified optimum equation, KED was carried out. Cross validation was used to assess the performance of the interpolators, and among the windows tested for KED an optimum window was chosen for each class. A diagram of the methodology is shown in Figure 4.2. This shows the sequence of procedures applied to choose the optimum interpolator.

Ordinary Kriging was also applied using a climatological variogram (OK-CV) obtained by pooling the datasets for each class. Developing a climatological variogram for Kriging with External Drift (KED) cannot be achieved simply through pooling the data. In order to produce an average residual variogram to apply in Regression Kriging, once the optimum window was chosen, the datasets belonging to each class were standardized by subtracting the mean and dividing by the standard deviation of each dataset. The averages of the standardized values in the class were used in stepwise regression with the topographic parameters of the optimum window, obtaining an average trend surface for each class. This surface was then used in Regression Kriging. Regression Kriging using and average residual variogram is referred to as RK-AV.

The results from the different interpolators were compared by using statistics of the cross validation to identify an optimum interpolator for each class of daily data. For Ordinary Kriging and Kriging with external drift the software R with the library GSTAT [Pebesma and Wesseling, 1998] was used. Details of the method are described in the following sections.

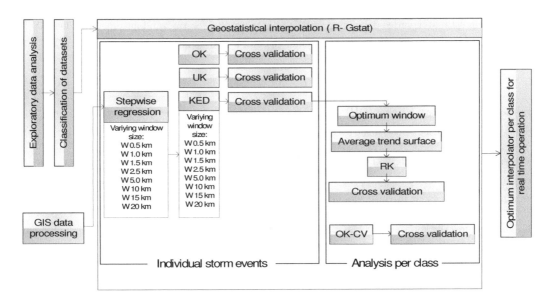

FIGURE 4.2: Methodology

4.2.3.1 Interpolation techniques

All Kriging estimators are variants of the basic linear regression estimator $Z^*(u)$ defined as [Goovaerts, 2000]:

$$Z^*(u) - m(u) = \sum_{\alpha=1}^{n(u)} \lambda_\alpha(u)[Z(u_\alpha) - m(u_\alpha)] \qquad (4.1)$$

Where u is an unsampled location in a study area A where there are n available data $z(u_\alpha), \alpha = 1, ..., n$, $\lambda_\alpha(u)$ is the weight assigned to datum $z(u_\alpha)$ interpreted as a realization of the random variable (RV) $Z(u_\alpha)$. The quantities $m(u)$ and $m(u_\alpha)$ are expected values of the RVs $Z(u)$ and $Z(u_\alpha)$ being $z(u)$ the realization of the unknown RV $Z(u)$. All types of Kriging share the same objective of minimizing the estimation or error variance $\sigma_E^2(u)$ under the constraint of zero bias of the estimator; this is:

$$\sigma_E^2(u) = Var[Z^*(u) - Z(u)] \qquad (4.2)$$

Is minimized under the constraint that:

$$E[Z^*(u) - Z(u)] = 0 \qquad (4.3)$$

The Kriging estimator varies depending on the adopted model for the random function. The RV is usually decomposed into a residual component $R(u)$ and a trend component $m(u)$:

$$Z(u) = R(u) + m(u) \qquad (4.4)$$

In the case of ordinary Kriging (OK) $m(u)$ accounts for local fluctuations of the mean by limiting the domain of stationarity of the mean to the local neighborhood.

When a drift is present Kriging with external drift can be applied. In this model the RV $Z(u)$ is considered as the sum of a deterministic drift (usually a polynomial with unknown coefficients) and a zero-mean stationary or intrinsic random residual $R(u)$ [Chiles and Delfiner, 1999].

Kriging requires the knowledge of the spatial structure of the random field in order to solve the Kriging system. This structure is characterized by a function called the variogram [Berne et al., 2004, Lebel and Bastin, 1989]:

$$\gamma(u_\alpha - u'_\alpha) = \frac{1}{2} E[Z(u_\alpha) - Z(u'_\alpha)]^2 \qquad (4.5)$$

Where u_α and u'_α are the position vectors. The experimental variogram can be inferred from spatially distributed measurements assuming that the expectation is equal to the arithmetic mean, and then fitted to a model (linear, spherical, exponential, circular, Gaussian, Bessel or power). Once the variogram model is estimated, this can be used to derive semivariances at all locations and solve the Kriging weights. The OK weights λ are solved by multiplying the covariance matrix C derived for $n \times n$ observations with the covariance vector c_0 at a new location:

$$\lambda = C^{-1}c_0 \qquad (4.6)$$

The lagrange multipliers are used in the matrixes to ensure that the sum of the weights is equal to 1. See Hengl [2009] for a complete description. In the case of KED and UK, the formulae are similar, except the covariance matrix and vector are extended with values of auxiliary predictors. For UK the only auxiliary predictors are the coordinates. On the other hand, in RK the predictions are made separately for the drift and the residuals and then added back together. This constitutes an advantage by allowing the use of arbitrarily complex forms of regression, rather than the simple linear techniques as can be used with KED. In addition, RK allows the separate interpretation of the two interpolated components [Hengl, 2009].

For OK, UK and KED data for each storm are used to derive the variogram. In the case of a climatological variogram, information of all realizations is taken into account, assuming the fields to have similar characteristics except for a constant value. The variogram can therefore be normalized by the respective variance of each field and then averaged over all the realizations [Berne et al., 2004].

The parametric structure of a climatological variogram is [Lebel and Bastin, 1985]:

$$\gamma(t_i, h) = \alpha(t_i) g(h, \beta) \tag{4.7}$$

Where h is the Euclidian distance, $\alpha(t_i)$ is a scaling parameter and β is a shape parameter. With this structure all the temporal nonstationarity (i.e. the dependence on the time index t_i) is concentrated in the scale factor $\alpha(t_i)$ (which has to be estimated separately for each event), while the factor $g(h, \beta)$ is time invariant and can be estimated once and for all from the complete set of data [Lebel and Bastin, 1985].

The parameter, $\alpha(t_i)$ then mainly accounts for the scale effect due to the variation in time of the mean rainfall intensity. $\alpha(t_i)$ being the variance of the (t_i) field and $g(h, \beta)$ the unique variogram of all the scaled random fields defined by:

$$z(u_\alpha)_{t_i}^{scaled} = \frac{z(u_\alpha)_{t_i}}{\sqrt{\alpha(t_i)}} \tag{4.8}$$

As can be expected from the normalization, the sill of the experimental variogram is nearly one [Lebel and Bastin, 1985].

4.2.3.2   Topographic parameters as secondary variables

The relation between precipitation and secondary variables extracted from topography is complex. The main variable that has been used is elevation. However, a simple correlation between precipitation and elevation does not always hold, and in several studies it has been found that the optimal correlation with elevation is not necessarily the point measurement but more often is the effective elevation of a larger area (called the window) surrounding the observation point [Daly et al., 1993, Kyriakidis et al., 2001]. Furthermore, several authors have found that smoothed topographic features in windows in the range 1-15 km show a high correlation with precipitation [Diodato and Ceccarelli, 2005, Hutchinson, 1998]. This procedure allows to smooth the local effects and a better integration of the main characteristics of the topographical environment [Portalés et al., 2008]. To identify an optimum size, squared windows of 0.5, 1, 2.5, 5, 10, 15 and 20 km were used to smooth the secondary variables listed in Table 4.1. The square shape was used mainly for reasons of computational efficiency [Isaaks and Srivastava, 1989] and the secondary variables were selected according to their relevance as presented in the literature.

To process the variables, a digital elevation model (DEM) with a pixel size of 30 meters was used. The statistical method used to relate precipitation to topographic variables was multiple linear regression and a stepwise approach was applied.

TABLE 4.1: Secondary variables for KED

Location	Abbreviation	Description
Location	Easting: s1 Northing: s2	Coordinates according to Bogotá coordinate system [Prudhomme, 1999]
Smoothed Elevation	Elevation	Smoothed elevation according to window size [Diodato, 2005]
Point Elevation	DTM	Elevation at station or at the centre of pixel [Hay et al., 1998]
Elevation Range	Elevation Range	Difference between highest and lowest elevation in a window [Hay et al., 1998]
Maximum elevation within wedge	Mxw_xx_yy	Maximum elevation within a wedge of given orientation (angle between xx and yy) and radius equal to the window size. [Agnew and Palutikof, 1999]
Aspect	Aspect	The values correspond to the compass direction of the downslope direction of the maximum rate of change in value from each cell to its neighbours, in a window [Buytaert et al., 2006b]
Slope	Slope	Slope of a window [Vidal and Varas, 1982]
Eastern and northern components	Eastern: p Northern: q	Eastern and northern components of the unit normal vector of the smoothed DEM. These variables permit the incorporation of the effects of both slope and aspect in a process oriented fashion [Hutchinson, 1998]
Maximum elevation	MAX	Maximum elevation in a window [Hay et al., 1998]

4.2.3.3 Cross validation and statistical criteria of comparison

In order to assess the performance of the interpolators, cross validation was used. This consists in temporarily discarding a sample value at a particular location from the sample data set; the value at that location is then estimated using the remaining samples [Isaaks and Srivastava, 1989]. The difference between the estimated value and the corresponding measured value is the experimental error ε [Diodato, 2005].

Several error measurements have been proposed to be used in cross validation. The mean error (ME) is used for determining the degree of bias in the estimates [Haberlandt, 2007], but it should be used cautiously because negative and positive estimates counteract each other and resultant ME tends to be lower than actual error. The root mean square error (RMSE) provides a measure of the error size [Diodato, 2005], but it is sensitive to outliers, whereas the mean absolute error (MAE) is less sensitive to extreme values. Models with a ME closer to 0 and a small RMSE are considered better.

The correlation between the observed values and predicted values, percentage of variance explained (PVE) or coefficient of determination is also a commonly used performance measurement.

TABLE 4.2: Descriptive statistics of the 139 daily data sets

Statistics	Max	Mean	Min
No of stations	85	64	39
Mean	32.7	12.4	2.54
Maximum	139	47.1	20.4
Median	29.1	9.82	0
Minimum	6.3	0.24	0
Standard Deviation	28.2	10.2	4.42
Variance	794.4	114	19.5
Skewness	4.14	1.33	-0.79
CV	3.46	0.96	0.3

4.2.3.4 Conditional Simulations

The influence of the selection of interpolator on the precipitation volume in the study area is estimated through conditional simulation. A conditional simulation is a realization randomly selected from the subset of realizations that match the sample points. Equivalently, it is a realization of a random function with a conditional spatial distribution [Chiles and Delfiner, 1999]. The mean of a large number of independent conditional simulations at a given point converges to the Kriging estimate, and their variance tends to the Kriging variance. Conditional simulation is useful to obtain a qualitatively, realistic picture of the spatial variability, while quantitatively, it can be used to evaluate the impact of spatial uncertainty [Chiles and Delfiner, 1999].

Spatial uncertainty is modelled by generating multiple realizations of the joint distribution of attribute values in space, a process known as stochastic simulation. Then, a transfer function can be applied to the set of realizations, yielding a distribution of the response values [Goovaerts, 1997]. In deriving the distribution of precipitation volume, 500 realizations of the rainfall fields are generated through conditional simulation, and then sampled over the study area.

4.3 Results

4.3.1 Exploratory data analysis

The descriptive statistics of the 139 rainfall fields that were analysed are shown in Table 4.2. This shows the statistics for the individual rain gauges.

Following selection of the storm and subsequent validation of data, it was found that for some storms the highest recorded rainfall was below the 30 millimetre threshold used to identify the rainfall events; however, it was decided to keep them in the analysis as these events were still considered significant.

TABLE 4.3: Occurrence of daily datasets per month

Month	EH-LE	EH-SE	FL-LE	FL-SE	FU-LE	FU-SE	MLT	UT
Jan	2	2	0	0	0	3	1	0
Feb	3	3	0	0	3	6	0	0
Mar	5	1	4	2	3	0	1	0
Apr	9	0	6	0	6	0	0	2
May	4	1	6	1	3	2	0	4
Jun	0	0	4	0	2	2	0	1
Jul	0	0	1	0	0	0	0	2
Aug	0	0	0	0	3	0	0	0
Sep	1	0	0	0	1	0	2	2
Oct	3	0	0	0	1	1	0	0
Nov	2	3	3	1	5	3	6	1
Dec	2	2	2	0	0	0	0	0
Total	31	12	26	4	27	17	10	12

The months with most daily data sets are April, May and November, which correspond to the rainy season. However, there are datasets in all months of the year. The statistics of the data sets show the variance and coefficients of variation to be high for most data sets. Commonly the highest values of precipitation in the data set are concentrated on small areas, reflecting the convective nature of precipitation in the area.

4.3.2   Classification of daily datasets

The occurrence of daily datasets according to class per month is shown in Table 4.3. Most of the daily datasets show precipitation occurring in the zones Eastern hills (EH) and Fucha (FU) with large and small extent. April is the month when most of the large extent storms take place while small extent storms concentrate during the first dry season of the year (January-February), which can be explained by the prevalence of convection. During June and the second dry season corresponding to July and August, some large extent events occur in the Flat (FL), Fucha (FU)and Upper Tunjuelo (UT) zones and small extent events occur during this period in the Fucha (FU) zone. During the second rainy season (from October to November) most of the chosen events take place in November in the middle and lower basin of the Tunjuelo (MLT) river as well as in the FU zone.

4.3.3   Variogram analysis

For Ordinary Kriging and Kriging with External Drift omnidirectional variograms were estimated for daily precipitation and residuals respectively, and models (Gaussian, exponential or spherical) were fitted using automatic fitting of GSTAT [Pebesma and Wesseling, 1998]. The median value of the nugget, sill and range of the variograms for each class and all datasets are shown in Table 4.4. For the

TABLE 4.4: Median values of Nugget, Sill and Range for the individual variograms fitted for the 139 datasets.

Class	Median value		
	Nugget	Sill	Range
All datasets	9	84	5951
Fucha (FU)	5	106	5922
Fucha Large Extent (FU-LE)	7	99	6291
Fucha Small Extent (FU-SE)	0	115	3965
Eastern Hills (EH)	12	97	5150
Eastern Hills Large Extent (EH-LE)	15	110	5841
Eastern Hills Small Extent (EH-SE)	10	95	3831
Flat (FL)	12	74	6238
Flat Large Extent (FL-LE)	12	75	6296
Flat Small Extent (FL-SE)	8	72	5756
Upper Tunjuelo (UT)	5	46	6269
Middle and Lower Tunjuelo (MLT)	14	54	9221

whole group of datasets the median correlation distance is about 6 km. The daily datasets belonging to the Fucha and Eastern Hills zones, that represent the storms that mainly take place in the eastern mountainous area exhibit the shortest range. The range of the small extent datasets is consistently shorter than the large extent datasets. The range for the Upper Tunjuelo and the Flat zone are similar. A division between large scale and small scale for zones UT and MLT was not made due to the small amount of datasets available.

Climatological variograms were constructed by pooling the datasets belonging to each class using equation 4.8, with results shown in Figure 4.3. In the case of the Upper Tunjuelo (UT), Middle and Lower Tunjuelo (MLT) and Flat (FL) zones, the climatological variogram to be used in the further analysis corresponds to the one obtained by pooling all the data of the zone, combining the large extent and small extent datasets. This is due to the availability of only two small extent datasets in the case of zone UT, and the insignificant difference between small and large extent variograms for zones MLT and FL.

In order to obtain an average residual variogram to be used in Regression Kriging, an average surface trend was fitted to each class. The datasets were standardized by subtracting the mean and dividing by the standard deviation and an average value for each rainfall station was obtained to be correlated with the secondary variables once an optimum window from the analysis of parameters of comparison of cross validation was carried out.

To apply regression the average residual variograms shown in Figure 4.4 were obtained. Using the average surface trend, the residuals for each dataset were calculated and divided by the standard deviation and subsequently pooled to obtain an average variogram. As expected the ranges are shorter than the ranges of the climatological variograms.

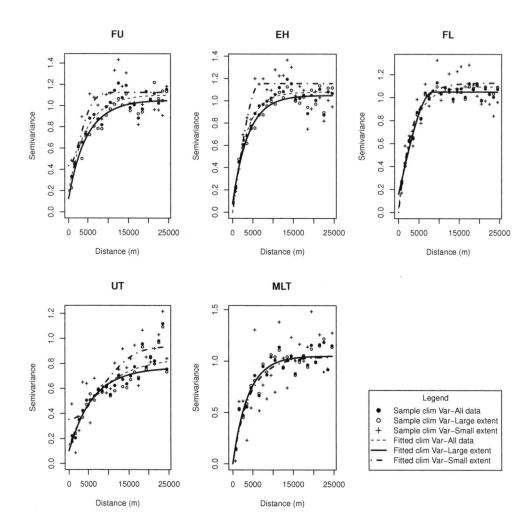

FIGURE 4.3: Climatological variograms for each zone

4.3.4   Analysis of performance of the interpolators for the individual storms

The results of cross validation of the rainfall fields applying ordinary Kriging and Kriging with external drift were compared for each dataset class.

The observed values where compared with the interpolated ones using the RMSE and PVE. This choice of cross-validation statistics was made taking into account that the PVE is correlated with the relative mean square error ReMSE [Syed et al., 2003] and RMSE is correlated with MAE. Additionally,

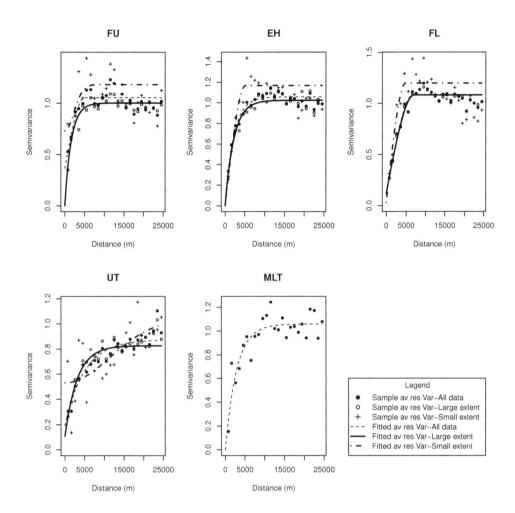

FIGURE 4.4: Average residual variograms for each class. No difference is made between large extent and small extent in the MLT zone due to low availability of data.

a good correlation between the RMSE and the maximum and the variance could be observed for all interpolators, indicating that the larger the maximum measured or the larger the variance, the larger the RMSE, meaning that the performance of the model is poorer in terms of size of the error.

The comparison of OK with KED for all the windows was carried out by comparing the statistical distribution of PVE and RMSE for each interpolator for each class. The results are shown in Table 4.5.

The second column of Table 4.5 shows the window with the highest adjusted R^2, which shows the best correlation between precipitation and secondary variables for each class, adjusted for the number

TABLE 4.5: Summary of results of comparison of interpolators

Class	Window Highest R^2	PVE Best Window	RMSE Best	Chosen Interpolator Window
FU-LE	15 km	10 km	10 km	10 km
FU-SE	5 km	10 km	10 km	10 km
EH-LE	2.5 km	5 km	5 km	5 km
EH-SE	1 km	1 km	2.5 km	2.5 km
FL	15 km	15 km	15 km	15 km
UT	15 km	10 km	10 km	10 km
MLT	10 km	10 km	10 km	10 km

of variables (p) used [Hengl, 2009]:

$$Adjusted \ R^2 = 1 - \frac{n-1}{n-p}(1 - R^2) \tag{4.9}$$

The behavior of adjusted R^2 is shown in Figure 4.5 for the Fucha class – large extent. Adjusted R^2 shows an increase from universal Kriging (using only coordinates as secondary variables) to Kriging with External Drift using a window of 1000 m. For larger windows the median values start to vary without a clear trend and an optimum window of 15000 m can be identified since the median, first and third quartile are the highest.

The behaviour of PVE for the Fucha class – large extent is shown in Figure 4.6. There is a significant change in the distribution of the PVE when KED with smoothed variables is used. In this class the window of 10000 m was identified as optimum since it has the highest first quartile, highest median and highest third quartile. Even though the median values and the first quartile of PVE increase noticeable when KED is applied, reduction in the upper whiskers takes place indicating deterioration in the performance of the interpolators in comparison with OK for some storms.

To compare the improvement of OK, in the case of RMSE, the RMSE/RMSE for OK is shown in Figure 4.7, this graph allows the visualization of the variation of RMSE using RMSE for OK as a reference. Despite most mean values being lower than 1 thus indicating an improvement, for some storms there is a deterioration as can be seen when the values are above 1.

From these graphs the optimum window can be determined where the mean is the lowest. This procedure was followed for all classes.

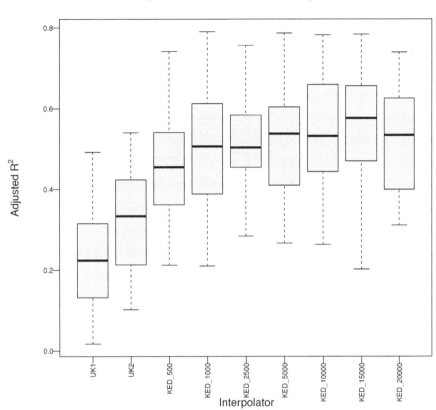

FIGURE 4.5: Adjusted R^2 for Fucha class – large extent datasets for all the interpolators. UK stands for Universal Kriging for a first order trend (UK1) and a second order trend (UK2), KED_xxxx stands for the interpolator Kriging with external drift with smoothed variables in a window of size xxxx [m].

4.3.5 Analysis of performance of the interpolators using the climatological variograms and applicability of the climatological variograms for individual event rainfall field generation

The comparison of OK, OK-CV, KED for the optimum window and RK-AV for each class is shown in Table 4.6. This table shows the minimum, first quartile, second quartile, third quartile and maximum of the PVE and RMSE relative to RMSE for OK (which was shown only for the Fucha class on Figures

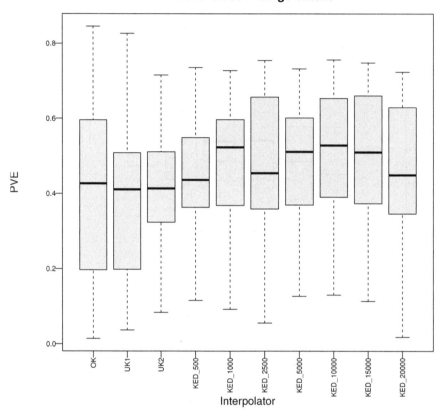

FIGURE 4.6: PVE for Fucha class – large extent datasets for all the interpolators. UK stands for Universal Kriging for a first order trend (UK1) and a second order trend (UK2), KED_xxxx stands for the interpolator Kriging with external drift with smoothed variables in a window of size xxxx [m].

4.6 and 4.7). The values in italics are the best overall and the bold values show the best interpolator using climatological variograms.

In the case of the Fucha datasets, KED shows good behaviour in terms of PVE in comparison with the other interpolators, however OK-CV shows very close values to KED and the behaviour of this interpolator in terms of RMSE is better since it presents the lowest third quartile and maximum. This means that it produces the least degradation in the behaviour in comparison with OK in the few cases where this deterioration takes place. Figure 4.8 shows the interpolated rainfall fields for the daily precipitation on 1st of April 2009, which was classified as FU-LE. The rainfall distribution

TABLE 4.6: Comparison of PVE and RMSE/RMSE for Ordinary Kriging for all the classes

	PVE						RMSE/RMSE for Ordinary Kriging					
	MIN	1Q	2Q	3Q	MAX	IQR	MIN	1Q	2Q	3Q	MAX	IQR
Fucha												
OK	-0.02	0.28	0.43	0.62	0.85	0.34	1.00	1.00	1.00	1.00	1.00	0.00
KED	*0.13*	*0.42*	0.50	0.65	*0.83*	*0.23*	*0.76*	*0.87*	*0.93*	1.06	1.33	0.20
OK-CV	**0.10**	**0.35**	*0.53*	*0.67*	**0.82**	**0.31**	**0.78**	**0.92**	**0.97**	*1.01*	*1.15*	*0.09*
RK-AV	0.12	0.31	0.46	0.61	0.76	0.30	0.78	0.95	1.00	1.07	1.21	0.12
Fucha Large Extent												
OK	0.01	0.21	0.42	0.59	0.85	0.38	1.00	1.00	1.00	1.00	1.00	0.00
KED	*0.13*	*0.37*	*0.51*	*0.65*	0.76	*0.28*	*0.76*	*0.87*	*0.93*	1.03	1.17	0.20
OK-CV	**0.10**	**0.32**	**0.46**	**0.64**	*0.82*	**0.32**	**0.92**	**0.95**	**0.98**	*1.01*	*1.05*	*0.09*
RK-AV	0.12	0.27	0.44	0.61	0.75	0.34	0.84	0.96	1.00	1.05	1.19	0.12
Fucha Small Extent												
OK	-0.02	0.38	0.48	0.68	0.84	0.29	1.00	1.00	1.00	1.00	1.00	0.00
KED	*0.42*	0.43	0.47	0.55	0.71	*0.12*	0.81	*0.86*	*0.93*	1.11	1.32	0.20
OK-CV	**0.28**	*0.44*	*0.58*	*0.71*	*0.80*	0.26	*0.76*	**0.87**	0.95	*1.02*	*1.07*	*0.09*
RK-AV	0.29	0.42	0.50	0.56	0.76	0.13	0.77	0.89	0.98	1.11	1.21	0.12
Eastern Hills												
OK	0.00	0.22	0.39	0.54	0.86	0.32	1.00	1.00	1.00	1.00	1.00	0.00
KED	-0.02	0.27	0.40	*0.56*	*0.83*	0.29	*0.75*	*0.92*	*0.99*	1.09	1.31	0.17
OK-CV	*0.05*	*0.30*	*0.43*	0.51	0.77	*0.21*	**0.86**	**0.96**	1.01	*1.04*	*1.15*	*0.09*
RK-AV	-0.02	0.29	0.40	0.53	0.78	0.24	0.82	0.95	1.02	1.08	1.16	0.14
Eastern Hills Large Extent												
OK	0.00	0.19	0.38	0.52	0.86	0.33	1.00	1.00	1.00	1.00	1.00	0.00
KED	0.11	*0.33*	*0.43*	0.50	0.76	*0.17*	*0.81*	*0.91*	*0.95*	1.04	1.13	0.13
OK-CV	*0.11*	**0.30**	**0.42**	*0.50*	*0.77*	0.20	**0.86**	**0.96**	1.01	*1.04*	*1.10*	*0.08*
RK-AV	-0.02	0.27	0.38	0.47	0.73	0.20	0.85	0.98	1.02	1.07	1.12	0.10
Eastern Hills Small Extent												
OK	0.07	0.35	0.52	0.55	0.67	0.20	1.00	1.00	1.00	1.00	1.00	0.00
KED	0.18	0.27	0.46	0.57	*0.64*	0.31	*0.77*	0.95	1.00	1.12	1.19	0.17
OK-CV	0.05	*0.34*	0.47	0.54	0.57	*0.20*	0.91	0.97	1.01	1.12	1.22	*0.15*
RK-AV	*0.19*	**0.33**	*0.50*	*0.59*	0.61	0.25	**0.92**	*0.92*	*0.99*	*1.09*	*1.16*	0.17
Flat												
OK	0.01	0.24	0.35	0.42	0.59	0.18	1.00	1.00	1.00	1.00	1.00	0.00
KED	*0.08*	*0.30*	*0.40*	*0.47*	*0.62*	0.18	*0.85*	*0.92*	*0.97*	1.03	1.15	0.11
OK-CV	**0.07**	**0.29**	**0.35**	**0.45**	**0.59**	*0.16*	**0.92**	**0.97**	1.00	*1.02*	*1.09*	*0.05*
RK-AV	0.02	0.25	0.37	0.45	0.57	0.19	0.89	0.96	1.00	1.03	1.10	0.07
Upper Tunjuelo												
OK	0.00	0.04	0.20	0.39	0.88	0.35	1.00	1.00	1.00	1.00	1.00	0.00
KED	*0.11*	*0.23*	0.27	*0.53*	*0.89*	*0.30*	0.82	0.87	*0.93*	0.99	1.09	0.12
OK-CV	0.02	0.05	0.25	0.38	0.86	0.33	0.90	0.93	1.00	1.06	*1.07*	0.13
RK-AV	**0.07**	**0.17**	*0.30*	0.48	0.89	0.32	*0.82*	*0.87*	0.95	*0.97*	1.09	*0.11*
Middle and Lower Tunjuelo												
OK	0.23	0.33	0.39	0.46	0.51	0.14	1.00	1.00	1.00	1.00	1.00	0.00
KED	*0.32*	*0.38*	*0.44*	0.50	*0.65*	0.12	*0.84*	*0.91*	0.98	1.02	1.06	0.11
OK-CV	0.30	0.34	0.38	0.44	0.58	*0.10*	0.92	0.95	1.00	1.00	1.06	0.05
RK-AV	**0.18**	**0.33**	**0.42**	*0.50*	0.62	0.17	0.87	0.93	*0.95*	*0.98*	*1.04*	*0.04*

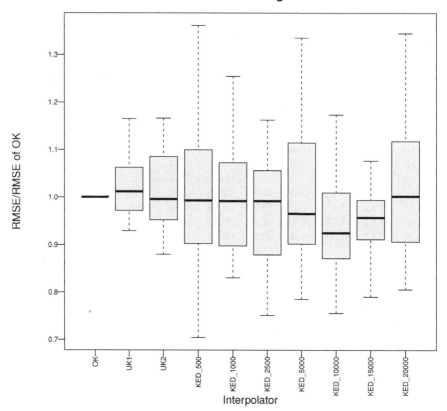

FIGURE 4.7: RMSE/RMSE of OK for Fucha class – large extent datasets for all the interpolators. UK stands for Universal Kriging for a first order trend (UK1) and a second order trend (UK2), KED_xxxx stands for the interpolator Kriging with external drift with smoothed variables in a window of size xxxx [m].

obtained from the interpolators OK and OK-CV provide a smoother surface in comparison with the results form KED and RK-AV where the rainfall field reflects the topographic characteristics of the area.

For EH datasets, OK-CV shows the best behaviour improving lower values of PVE and producing the least deterioration of RMSE for the case of all datasets. When large extent datasets are analysed, KED is very similar to OK-CV with OK-CV slightly superior since it produces the least deterioration of RMSE. In the case of small extent datasets for EH class, RK-AV shows a median value close to OK and a shorter range towards lower values of PVE, additionally it shows the lowest median of RMSE.

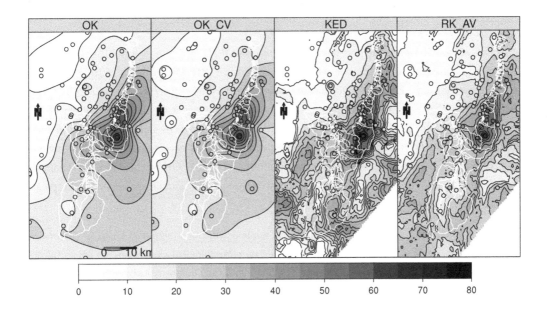

FIGURE 4.8: Rainfall fields for daily precipitation on 1st of April 2009. Precipitation according to the color key is given in millimeters.

In the case of FL datasets, in terms of PVE, the behaviour of KED, OK-CV and RK-AV are similar, with KED slightly superior. In terms of RMSE the lowest median in comparison with RMSE for ordinary Kriging corresponds to KED, however, OK-CV shows the shortest interquartile range and the lowest extreme value above 1, which means less possibility of deterioration of the performance in comparison with OK. The improvement obtained by using KED can be equalled by OK-CV leading to a simplification of the procedure. The same performance in terms of PVE can be expected from both interpolators and the RMSE is slightly higher for OK-CV but in the cases where this parameter deteriorates, a smaller deterioration can be expected from OK-CV than from KED.

In the case of the UT class, the best performance in PVE corresponds to KED and RK-AV. The interquartile range of both is similar but the median value of RK-AV is higher meaning a better performance of this interpolator. In terms of the RMSE compared with RMSE for ordinary Kriging the behaviour of KED and RK-AV is similar with a slightly higher median value for RK-AV, but both with the interquartile range under 1 meaning that a significant improvement of RMSE can be obtained by using KED. The difference between KED and RK-AV is not significant, so the topographic parameters can be incorporated in the interpolation by using an average residual variogram.

In the case of the MLT class the best behaviour in terms of PVE corresponds to KED with the highest first quartile and median, followed closely in performance by RK-AV. In terms of RMSE compared to RMSE for ordinary Kriging, a significant improvement can be observed for KED, OK-CV and RK-AV. The best performance corresponds to RK-AV with the lowest median value and lowest third quartile significantly less than 1 which means an improvement over OK in most cases and a smaller interquartile range. For this class of datasets, the behaviour of OK is very similar to the behaviour of OK-CV with even some improvement of the latter in terms of RMSE. OK can be improved significantly by KED and RK-AV. The latter behaves better than KED, allowing simplification of the interpolation procedure and improving performance.

This means that a climatological method can be applied without a significant loss in performance given that in the cases when the interpolation based on climatological variograms do not outperform the other methods, the decrease of performance when using the climatological variograms in terms of the median value of the distribution of PVE reaches a maximum of only 13%.

4.3.6 Analysis of secondary variables

An analysis of the secondary variables used in KED was carried out, aimed at identifying their importance in the interpolation of the rainfall field. Stepwise regression between precipitation and secondary variables was applied, first to all the datasets individually and afterwards to the averaged standardized precipitation, to obtain a unique regression equation for the optimum window.

To establish the relative importance of the secondary variables, the standardized regression coefficients or beta weights were calculated as the standard deviation change in the dependent variable when the independent variable is changed by one standard deviation [Bring, 1994]. The results of the analysis of standardized beta weights of the unique regression equation for the standardized average precipitation are shown in Figure 4.9. The unique regression equations show the importance of the east and north coordinates which are present in all the equations. Likewise, the maximum elevation in a wedge with a radius equal to the window size is present in all regressions (except for the class Fucha-small extent). This suggests the importance of the orientation of the mountains in relation to the wind. The ranking of the other parameters, however, varies for each class.

Class	Adjusted R^2	Secondary variables								
EH-LE	0.43	s1 * s2	s2^2	mx-w-90-180	mx-w-0-90					
EH-SE	0.46	mx-w-90-180	MAX	s1 * s2	Slope	s2^2				
FL	0.42	s1	s1^2	Aspect	q	MAX	mx-w-270-360			
FU-LE	0.76	s1^2	DTM	mx-w-90-180	Aspect	s1 * s2	Elevation	mx-w-180-270		
FU-SE	0.58	s1	s1^2	Slope	p	q	s2	DTM		
MLT	0.44	s1	s1^2	Aspect	mx-w-90-180	mx-w-0-90				
UT	0.49	s2^2	s2	s1 * s2	s1	Slope	mx-w-0-90	mx-w-270-360	q	DTM

Higher importance   Lower importance

FIGURE 4.9: Secondary variables in the unique regression equations for the standardized average precipitation. The notation of the variables is as follows: s1=easting, s2=northing, mx-w-xx-yy= maximum elevation within a wedge with orientation between xx and yy degrees calculated for the optimum window, MAX=maximum elevation in the optimum window, DTM= elevation at station, p and q=eastern and northern components of the unit normal vector to the smoothed DEM for the optimum window.

The results obtained for individual regressions for each daily dataset belonging to the FU-LE class are shown in Figure 4.10 as an example of the results obtained for all classes. The name of each column corresponds to the order of significance of each variable in the regression, the number in each cell indicates the number of times the variable appears in a given significance rank, with the variables with the darker shading being the most important. The last column shows the total number of times a variable appears in the regression equations.

Individually, the FU-LE datasets show a high significance of the east and north coordinates in the regressions, being the four most important variables for all the regression equations, which is consistent with the unique regression equation, where $s1^2$ and $s1*s2$ were included (see Figure 4.9). The variable that appears the least in the regression equations is max-w-0-90 as it appears only 8 times and the least important is max-w-270-360 because when it appears in the equations it has a low rank according to Figure 4.10. The variables max-w-90-180 and max-w-180-270 are identified as important both in the individual regressions and in the unique regression shown in Figure 4.9, which can be interpreted as the identification of the importance of the wind direction. Other variables such as p, slope and aspect are identified as important in Figure 4.10. However, the unique regression equation shows only aspect as being important. The unique regression seems to identify in a clearer way the most important variables than the analysis of standardized beta weights summarized in Figure 4.10.

4.3.7   Analysis of uncertainty in estimates of storm volumes

The results of the Gaussian simulations are shown in Figure 4.11. This shows the volume of daily precipitation for the Eastern Hills small extent storms, similar results were found in the other zones. This boxplot shows the interquartile ranges and median values for 500 daily volumes calculated from

Variable	Rank																	Total

Variable	1	2	3	4	5	6	7	8	9	10	11	12	13	14	15	16	17	Total
s1 * s2	8	1	3	3	0	2	0	1	0	0	0	0	0	0	0	0	0	18
s1	6	4	4	2	1	0	0	1	0	0	0	0	0	0	0	0	0	18
s2	5	6	1	1	1	0	0	1	0	0	0	0	0	0	0	0	0	15
s1^2	2	5	2	0	0	0	0	0	0	0	0	1	0	0	0	0	0	10
p	2	1	3	2	2	2	2	0	0	0	0	0	0	0	0	0	0	14
Slope	1	2	0	3	3	2	0	0	1	1	0	0	0	0	0	0	0	13
Aspect	1	0	2	3	2	1	2	0	1	0	0	0	0	0	0	0	0	12
mx-w-90-180	1	0	2	0	2	2	0	0	1	1	1	0	1	0	0	0	0	11
mx-w-180-270	1	0	0	0	4	1	3	2	2	2	0	0	0	0	0	0	0	15
s2^2	0	4	4	2	0	0	0	0	0	0	0	0	0	0	0	0	0	10
q	0	3	0	5	2	2	1	1	2	1	0	0	0	0	0	0	0	17
Elevation Range	0	1	1	1	2	2	2	3	0	0	0	0	0	0	0	0	0	12
Elevation	0	0	3	0	1	3	4	3	0	0	0	0	0	0	0	0	0	14
MAX	0	0	1	2	4	0	0	2	1	1	0	0	0	0	0	0	0	11
mx-w-0-90	0	0	0	2	0	0	2	0	2	0	1	1	0	0	0	0	0	8
DTM	0	0	0	0	1	3	3	1	1	1	4	0	0	0	0	0	0	14
mx-w-270-360	0	0	0	0	1	1	0	2	4	1	0	1	0	0	0	0	0	10

FIGURE 4.10: Ranking of secondary variables according to significance in the individual regression equations for class FU-LE

the gaussian conditional simulations carried out with the four interpolators (OK, KED, OK-CV and RK-AV).

The comparison of interquartile ranges indicates a similar variation in the precipitation volume for all the interpolators. However, differences in the median values can be observed but there is no indication of a consistent bias among interpolators.

4.4 Discussion

4.4.1 Characteristics of the rainfall fields

The characteristics of the storms that occur in the study area pose a challenge in the search of an interpolation procedure. These storms present a high variability and the concentration of high values of precipitation over small areas which create difficulty in the determination of the spatial structure using the available rainfall network.

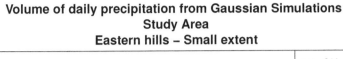

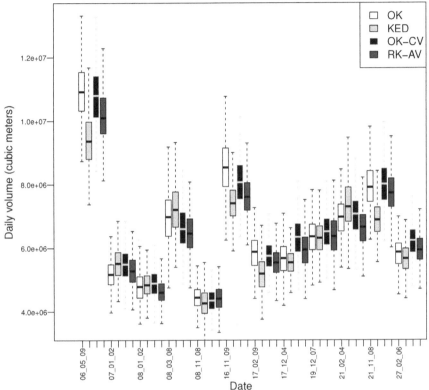

FIGURE 4.11: Comparison of simulated daily precipitation volumes for OK, KED, OK-CV and RK-AV for the Eastern hills – small extent class.

The classification of daily datasets shows typical patterns of occurrence of high precipitation storms, being April, May and November the months during which most of the chosen daily datasets take place. However, from the statistical analysis of the variograms it was found that it was not possible to identify a seasonal pattern that could explain the variability in the parameters of these variograms, as the sill and range do not vary consistently with the time of the year. This apparent lack of seasonality may be due to the inherently large variability of precipitation in the tropical zone, influenced by synoptic, mesoscale and microscale systems throughout the year [Bernal et al., 2007]. As a result, only the classification based on extent and location of the maximum of the daily data set was used. However, in other areas where rainfall is strongly seasonal a further classification of rainfall events may be needed to improve the climatological variograms. Daily rainfall can show a strong seasonality as found by Van de Beek et al. [2011] in The Netherlands. However, for the same region Schuurmans et al. [2007] did not find a clear seasonal effect using a different set of data. These results show the complexity in identifying seasonal patterns, even in mid-latitude regions.

The analysis of the individual variograms showed a short range (a median value of 6 km). This is comparable to ranges found in other regions of the Andes, as reported by Buytaert et al. [2006b] for mountainous areas of Ecuador, where a range of 4 km was found. On the other hand, the 6 kilometer median range contrasts with ranges up to 200 km that have been found in other areas of the world like the Netherlands [Van de Beek et al., 2011]. Given the complexity of the topography in the Bogotá area and that the rainfall is convective, such short ranges could be expected.

4.4.2 Performance of the climatological variograms and applicability to the generation of individual event rainfall fields

When the climatological variograms for each class were derived, a smoother and more robust spatial structure was found. The same occurred for the average residual variogram, which was constructed under the assumption of an underlying surface trend that can explain part of the variability of the daily datasets and that is common to all datasets belonging to the same class.

The analysis of the KED with the chosen window, OK, OK-CV and RK-AV showed that in most cases OK-CV or RK-AV can be used to carry out the interpolation, obtaining an improvement over OK. However, where OK shows good performance it is difficult to improve significantly on this using any of the more complex interpolators, with these even showing a deterioration in performance for some storms. In the cases where OK-CV provides the best results, showing a better performance when OK is very poor, and producing the least degradation of performance when that of OK is very good. In three cases RK-AV was considered the best interpolator with a similar or even better performance than KED. In the condition that OK-CV provides better performance than RK-AV the autocorrelation of the precipitation data that can be represented by the spatial structure as defined by the climatological variogram is more important than the correlation of precipitation with the secondary variables as expressed by the adjusted R^2. Where RK-AV provides better performance than OK-CV, the opposite occurs. This means that both interpolation techniques using OK-CV and RK-AV are promising for real-time operation as these provide comparable performance to using variograms derived for the individual storms (as in OK and KED). The choice between RK-AV and OK-CV can be based on a comparison of the importance of the correlation of precipitation with the secondary variables, when compared to the autocorrelation of the data. The use of spatial structures previously determined allows for automatic application of Kriging, which is prerequisite for real-time operation.

Individual rainfall fields can be generated through a procedure that in real-time identifies the appropriate class by determining the location of the maximum precipitation and the extent. According to the classification, the associated interpolation method can be chosen (RK-AV or OK-CV), for which climatological variograms are pre-defined as well as average trend surfaces in the case where RK-AV applies. Thus, automatic interpolation can be carried out based on climatological variograms that involve data from all similar previous events at daily scale. However, smaller time steps may be needed for flood early warning systems in the study area given the rapid hydrologic response of the watersheds. The sill and range of the variogram models are scale dependent, the scale being expressed by the rainfall duration [Bargaoui and Chebbi, 2009]. Lebel et al. [1987] and Berne et al. [2004] found a power relationship between the rainfall duration and the range of the climatological variogram for durations ranging between 1 minute and 24 hours. However, fitting the variogram for time steps smaller than 24 hours presents the difficulty of considerable scatter for sparse rain gauges with high resolution as in the study area. Therefore it is expected that a proper variogram structure for daily precipitation provides a base for the interpolation at smaller time steps. Furthermore, the relation between topography and rainfall is not so obvious for rainfall intensities of short duration [Bargaoui and Chebbi, 2009], which means that the study of the importance of secondary variables may not be feasible at sub-daily time scales.

For interpolation in real time, time steps smaller than 24 hours will be needed, which would mean smaller ranges that may not be properly captured by the sparse network in the study area. One option is to neglect the scale dependence of the range of the variogram and use the daily variogram. The influence of this approach would need further investigation.

Further merging with satellite and radar products may provide improvement in rainfall field estimations at these shorter timescales. Geostatistical techniques provide a means to combine both sources of data and has shown promising results [Grimes et al., 1999]. Projects such as the Global Precipitation Measurement (GPM) mission will provide high resolution and frequent observation of precipitation at global scale. However, one of the challenges is to produce precipitation estimates combining several sources (satellites, gauges and radars where available) that are significantly different in scale and resolution.

4.4.3 Choice between KED and OK

The comparison of PVE and RMSE for OK, UK and KED for the chosen windows showed an increase in adjusted R^2 when the smoothed secondary variables were used, and it was possible to identify an optimum window for each class by analyzing the distribution of PVE and RMSE. Even if an optimum window was chosen, there is, however, in some cases a deterioration in the behaviour when compared to OK, as represented by the increase of RMSE and the reduction of PVE. In 18 of 139 rainfall fields a deterioration of PVE occurred for all the tested windows in KED. This behaviour can be explained in the light of Figure 4.12. This shows the adjusted R^2 between precipitation and secondary variables of the best interpolator compared to the PVE found using OK. The 18 datasets that show deterioration correspond mainly to high performance of OK in terms of PVE and the adjusted R^2 that would be needed to produce improvement is not achieved by the secondary variables. This means that the adjusted R^2 needs to be higher than the PVE in order to obtain improvement when KED is applied. Where this is not the case, the results using OK would seem the most robust, as introducing the secondary variables results in a deterioration of performance. This shows that the choice between OK and KED should be done carefully, since the performance of KED can be lower when the correlation between the secondary variables and precipitation is smaller than the percentage of variability explained found in Ordinary Kriging.

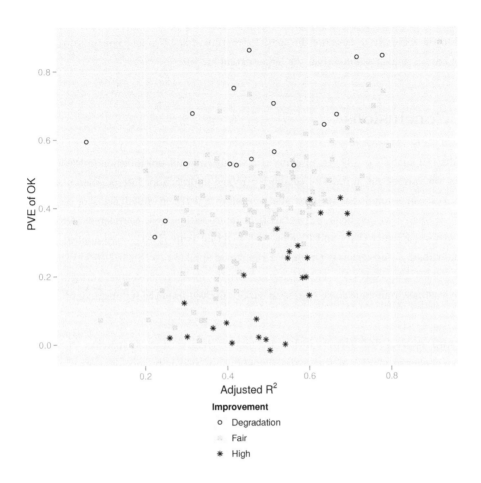

FIGURE 4.12: Improvement according to the relationship PVE of OK and adjusted R^2

4.4.4 Volumetric comparison

A comparison of the interquartile ranges of the daily volume obtained from the Gaussian simulations shows similarity among interpolators, with typical interquartile ranges between 5 and 20% of the magnitude of the storm expressed as the mean value of the simulated volumes. However, this variation of interquartile ranges differs depending on the zone and storm magnitude. For the Upper Tunjuelo the ranges are smaller. This could be attributed to the rainfall stations in this area being more evenly distributed, thus reducing the uncertainty in the volume estimate of the storm.

There is some correlation between the interquartile range and the magnitude of the storm. The correlation is negative, which means that the relative error is larger for smaller storms. However, R^2 is 0.33 for all the storms which means that the correlation may not be significant. On the other hand, this behaviour would be expected given that a higher uncertainty in the precipitation volumes is linked to the difficulty to measure the spatial structure of small extent precipitation events, which generally have a smaller volume than the larger extent storms.

4.5 Conclusions

A geostatistical analysis of 139 sets of daily data with significant precipitation in the area of Bogotá, Colombia was carried out in order to establish a robust approach for interpolation of rainfall, where that approach can be applied as an automated procedure in deriving rainfall fields for use in operational flood forecasting. Five methods of geostatistical interpolation were considered. Ordinary Kriging (OK), Universal Kriging (UK) and Kriging with external Drift (KED) were applied to each of the storms individually. Subsequently Ordinary Kriging was applied to a climatological variogram derived from the pooled data (OK-CV). For Regression Kriging an average residual variogram was derived using the residuals from an average regression surface derived from the set of standardized storm data (RK-AV).

Kriging with external drift (KED) with individual residual variograms increases the percentage of variability explained (PVE) of the spatial variability of the daily precipitation in most datasets. Using smoothed secondary variables, the results show an improvement over those found with Ordinary Kriging in most cases. However, the amount of improvement is shown to depend on the relationship between the PVE of Ordinary Kriging and the adjusted R^2 of the correlation of precipitation and the smoothed secondary variables. Only when the adjusted R^2 is significantly higher than the PEV of OK will a significant improvement be obtained, and where this is not the case considering secondary variables may even be detrimental. Due to this, no interpolator based on KED can be said to be better than OK in all cases. Therefore, the use of secondary variables and the inherent complexity in the procedure should only be considered when the performance or ordinary Kriging is poor in comparison with the correlation of precipitation and secondary variables.

The differences in performance between individual variogram interpolation (OK and KED) and pooled variogram interpolation (OK-CV and RK-AV) are not significant for the analysed storms, implying that the simplified, automatic procedures can be implemented without significant loss of performance for flood early warning purposes. Climatological variograms derived using pooled data were shown to be robust. An average trend for the storms belonging to a class could be defined allowing the construction of an average variogram of residuals. This variogram could be used successfully to interpolate the rainfall field without significant loss of performance and in some zones in the study area where the spatial structure of the precipitation data is poor, but the interpolation can be improved by using the average residual variogram, even showed to be superior. On the other hand, the superior performance of OK-CV can be explained by less degradation or even improvement when the performance of OK is high (contrary to the behavior of KED), accounting for a better spatial structure definition.

The above means that a smoother unique spatial structure of the variogram is able to improve the interpolation further than the topographic parameters when the correlation of the latter with precipitation is not sufficiently high. In these cases the best performance is found with OK-CV. And in the case of RK-AV showing the best performance, the correlation of secondary variables with precipitation is able to compensate the lack of structure in the data contributing to improve the rainfall field estimation.

When the volume of precipitation obtained from Gaussian simulations is analysed, no reduction in the variability can be associated to any interpolator. Likewise, no indication of a consistent bias among interpolators could be identified. Therefore, the choice of the preferred interpolator primarily depends on the analysis of PVE and RMSE.

Results show that Ordinary Kriging using a climatological variogram as well as Regression Kriging based on an average residual variogram provide robust techniques to obtain rainfall fields in real-time operation for flood early warning purposes. Thus, allowing the determination of a spatial structure previously to the occurrence of precipitation, incorporating all the spatial information of antecedent storms overcoming potential lack of enough rainfall stations during operation for the definition of the spatial structure and providing a time efficient automated procedure.

Chapter 5

Hydrological model assessment for flood early warning in a tropical high mountain basin

This chapter is an edited version of: Rogelis, M. C., Werner, M., Obregón, N., and Wright, N.: Hydrological model assessment for flood early warning in a tropical high mountain basin, Hydrol. Earth Syst. Sci. Discuss., doi:10.5194/hess-2016-30, in review, 2016.

5.1 Introduction

Models constitute the heart of early warning systems, providing a description of the hazard and its evolution in time [Basher, 2006]. Hydrologic and hydrodynamic models with varying levels of complexity are used to provide advance warning of the likely timing and magnitude of flooding, and to help to understand the complexities of a flood event as it develops [Sene, 2008]. A key aspect is to ensure that all relevant hydrological processes are included, and that appropriate computational weight is given to each process on the basis of its relative importance [Clark et al., 2008]. This task is highly complex, since different models represent hydrologic processes differently, and all of them are imperfect [Duan et al., 1992].

Hydrologic modelling is affected by four main sources of uncertainty: input uncertainty, output uncertainty, structural uncertainty and parametric uncertainty [Renard et al., 2010]. Structural uncertainty is defined as the modelling uncertainty due to the selection of an appropriate model, which includes the defined hydrological processes (perceptual model) and description of these processes (conceptual model) [Zhang et al., 2011], and their mathematical implementation. Uncertainty induced by model structures can be more significant than parameter and input data uncertainty, but such uncertainties are difficult to assess explicitly or to separate from other uncertainties during the calibration process [Beven and Binley, 1992]. The identification of the most appropriate model and model structure and its associated uncertainty to be implemented in a flood forecasting system is crucial, since the acceptable reproduction of hydrological processes builds up reliability into the hydrological model. This is essential when the model is to be used for forecasting and extrapolation [Reusser, 2010], where getting the "right answers for the right reasons" [Kirchner, 2006] or realism [Kavetski and Fenicia, 2011] is an important component of the confidence of the forecasting system. However the range of schemes available for assessing the impact of model structures on modelling uncertainty is still quite limited [Zhang et al., 2011].

The suitability of a rainfall-runoff model structure for a certain catchment has recently been studied through the use of flexible hydrological model structures, which focus on the diagnosis of their differences [Clark et al., 2008]. These flexible hydrological model structures include: the Framework for Understanding Structural Errors (FUSE) introduced by Clark et al. [2008]; the SUPERFLEX modelling framework proposed by Fenicia et al. [2011] that develops the earlier FLEX model [Fenicia et al., 2008]; and the Framework for Assessing the Realism of Model Structures (FARM) proposed by Euser et al. [2013], where consistency and performance are analysed through principal component analysis. The criteria to be used for model evaluation both in these frameworks and in standard calibration procedures are an active research topic. Metrics such as the Nash–Sutcliffe efficiency (NSE) [Nash and Sutcliffe, 1970] or the root mean square error (RMSE) are often used to evaluate simulation results. However, their drawbacks [Fenicia et al., 2007, Pfannerstill et al., 2014] call for a more comprehensive approach. The use of vector search techniques to optimize model parameters is an alternative to incorporate multiple criteria within calibration to provide a number of alternative parameter sets that are optimal, on the basis of the Pareto-dominance concept [Efstratiadis and Koutsoyiannis, 2010]. Fenicia et al. [2007] compared a pareto-optimality based calibration approach with a procedure that replicates the steps that are undertaken during manual calibration finding that given their strengths both calibration approaches can be combined. Other approaches rely on signature measures [Pfannerstill et al., 2014, Yilmaz et al., 2008] that define the hydrologic response characteristics and provide insight into the hydrologic function of catchments [Sawicz et al., 2011], rather than assessing model performance solely on the discharge at the outlet.

This study explores the suitability of three differing model concepts to be used for flood forecasting purposes in a basin located in Bogotá (Colombia). The aim of the research is to explore the performance of the models in order to identify the most appropriate modelling approach, given the characteristics of the study area. A lumped model (HECHMS Soil Moisture Accounting), a semi-distributed model (TOPMODEL) and a distributed model (TETIS) were used. In the case of the semi-distributed and distributed model, resolution was explored in order to identify the most suitable pixel size to be used. Finally, a comparison of precipitation input uncertainty and model performance is carried out in order to identify the importance of these in the modelling results, which constitutes relevant information for future improvement to the models.

The study area exhibits a high degree of complexity, since the upper basin corresponds to a páramo area (tropical high montane ecosystem), characterised by soils with a high water storage capacity and high conductivity with a hydrologic behaviour for which still major gaps in knowledge exist [Buytaert et al., 2006a, 2005b, Reyes, 2014, Sevink, 2007] and where the hydrometeorological data are scarce. Most modelling efforts in páramo areas have been carried out in micro-watersheds [Buytaert and Beven, 2011, Buytaert et al., 2004, 2007, 2006c, 2005b, Crespo et al., 2011] and have focused on advancing the understanding of hydrological processes and anthropogenic impacts. However, there is a relevant need to model larger páramo watersheds [Crespo et al., 2012], and advance in the challenge to produce forecasts for flood early warning to downstream communities. Previous modelling efforts include the use of the AvSWAT model [Díaz-Granados et al., 2005], the use of the linear reservoir model to study land-use changes [Buytaert et al., 2004], a combination of linear reservoirs and TOPMODEL to assess the hydrological functioning of the páramo ecosystem [Buytaert and Beven, 2011] and the analysis of climate change impacts through the use of the WEAP model [Vergara et al., 2011].

5.2   Study Area

Páramos constitute the source of water for Bogotá, the capital city of Colombia. Water is supplied by three main páramo systems namely Chingaza, Sumapaz and Tibitoc (Empresa de Acueducto y Alcantarillado de Bogotá 2015). The Tunjuelo river basin (see Figure 5.1) with an area of approximately 380 km^2, is located in the south of the city of Bogotá. The upper part of the basin is a páramo area where two reservoirs (Chisaca and Regadera) with volumes of 3.3 Mm^3 and 6.7 Mm^3 operate to supply 1.2 m^3/s of water to the south of Bogotá. This area belongs to the Sumapaz páramo, which is the largest páramo of the world [Daza et al., 2014]. It faces threats such as burning, inappropriate cropping, extensive cattle raising, mining, afforestation with inappropriate species, among others [Daza et al., 2014]. The middle basin corresponds to the transition from the rural area to the urban area of Bogotá (see Figure 5.1).

In 2006, a dry dam (Cantarrana Dam) was constructed in the middle basin for flood control purposes given the history of flooding of the Tunjuelo river (see Figure 5.1). The last significant flood occurred in 2002 causing the river to change its course, flowing into two mining pits that currently act as inline reservoirs. In the urban area three retention basins are located upstream of the confluence of the Tunjuelo river with the Bogotá river.

The watershed has a unimodal precipitation regime in the upper part (rainy season April-November) that transforms into a bimodal regime in the lower basin, with rainy seasons in March-May and

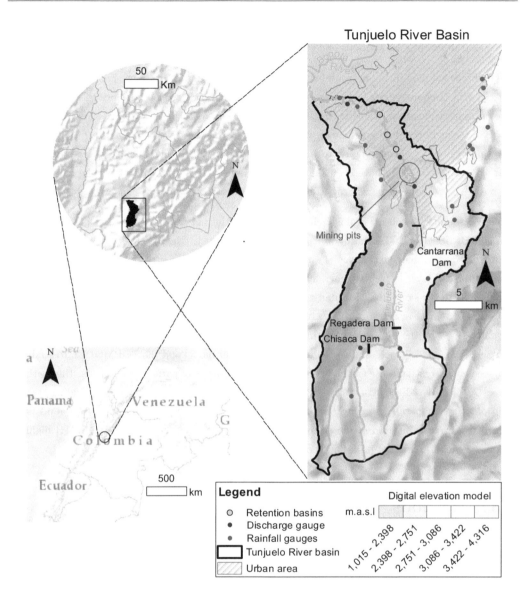

FIGURE 5.1: Study area. Service layer credits: Esri, DeLorme, USGS, NPS, USGS and NOAA

September-November. The average annual precipitation varies with the influence of the topography; from 600 mm in the North-West to 1500 mm in the upper basin (South-West) [Bernal et al., 2007].

The geology of the watershed consists of sedimentary rocks of Cretaceous, tertiary and quaternary age [INGETEC, 2002]. These sedimentary rocks form mountains up to 4000 m altitude, thus reaching some 1500 m above the level of the high altitude plain of Bogotá [Torres et al., 2005]. The main soils correspond to inceptisols, andisoils and entisoils (characteristic of páramo areas).

The hydrological monitoring network installed in the basin is shown in Figure 5.1. Although tipping bucket telemetric rain gauges have been operating in the Tunjuelo river basin since the year 2000, the development of the network has been gradual, and only in 2008 the network extended to cover the upper watershed. Six discharge gauges were selected in this analysis, three of which are located in the upper watershed. Rain gauges provide data each 10 minutes, while discharge gauges report each hour. Even though, there have been significant efforts in recent years to improve the monitoring of the basin, the monitoring network is still considered sparse.

Table 5.1 summarizes the hydrologic characteristics of the upper watershed of the Tunjuelo river. According to Sevink [2007] the hydrological processes in such páramo areas are fairly simple and dominated by two main pathways: (1) interflow through the upper litter layer, and (2) percolation through the soil layer (which is generally less than 1 m thick) down to the bedrock and subsurface flow parallel to the slope in a saturated zone just above the bedrock. As a result, simple models such as a set of two linear reservoirs already give satisfactory results. The characteristics of these reservoirs are determined by the flow velocities through the respective pathways. Buytaert et al. [2004] and Buytaert et al. [2005a] successfully used the linear reservoir model and the TOPMODEL to study the influence of different land use on the hydrological characteristics of páramo watersheds. Buytaert and Beven [2011] analysed the structure of 9 models to represent the páramo hydrology, finding that the addition of a slow parallel store to the original TOPMODEL [Beven and Kirkby, 1979] appears the most realistic representation of the system to date. However, Buytaert and Beven [2011] highlight that a correct estimation of peak flow remains a challenge.

5.3   Methods

The methodology is composed of three parts: model setup and calibration; performance analysis and diagnostics; and analysis of precipitation input uncertainty and comparison of models. Three model codes were selected; TETIS [Frances, 2012], HECHMSSMA (HEC HMS Soil Moisture Accounting) [USACE, 2000] and the TOPMODEL [Beven and Kirkby, 1979]. These were chosen based on previous use and identified to be suitable in mountainous and páramo areas for the case of TETIS and TOPMODEL [Sevink, 2007], and on the convenience of the HECHMS software since it is widely used in Colombia. However, criteria such as the simplicity and low computational demand were also taken into account. Initial parameters were derived from existing soil data and topography and calibration was carried out using the Shuffled Complex Evolution automatic search algorithm [Duan et al., 1992]. On the basis of the calibration results, a performance analysis and diagnosis of each model was carried out by using selected standard performance indices, as well as analysing how well the hydrological

TABLE 5.1: Hydrologic characteristics of the páramo area in the Tunjuelo river basin

Component	Description
Forcing data	Horizontal precipitation, fog and mist play an important role in the water balance [Díaz-granados et al., 2002]. There are, however, no measurements available for the study area. Rainfall events in the páramo are typically of high frequency and low intensity. In combination with strong winds and a very rough topography (rain shading) this results in high spatial rainfall variability and large errors in precipitation registration [Buytaert et al., 2004]. Actual evapotranspiration is low due to the presence of xerophytic plants [Buytaert et al., 2006c], low temperature, high frequency of fog, cloud cover and high relative humidity [Buytaert and Beven, 2011, Buytaert et al., 2011, Reyes, 2014]. Literature values of actual evapotranspiration range from 0.8 mm/day to about 1.5 mm/day [Buytaert et al., 2004, Hofstede et al., 1995].
Vegetation	Because of a predominance of grass species, water storage in the vegetation layer is minimal [Buytaert et al., 2005b]. However, natural páramo vegetation play an important role in the water cycle with a hydrologic behaviour that is as yet poorly understood [Buytaert et al., 2006a].
Soils	The soils in the páramo area correspond mostly to inceptisols, although andisols and entisols are present. These characteristic páramo soils have a high content of organic matter, high porosity, a large hydraulic conductivity [Buytaert et al., 2006a]. Infiltration capacities between 15 and 150 mm/h, and water retention capacities up to 90 vol% in saturated conditions [Buytaert and Beven, 2011]. Soils are relatively shallow (about 50 cm). The soils effectively regulate water producing a slow hydrologic response caused by the combination of a high water storage capacity and high conductivity [Buytaert et al., 2005a]. Thus the soil acts as a buffering reservoir, and turns the variable rainfall into a continuous water discharge [Buytaert et al., 2004]. Changes in soil moisture storage over time are relatively small [Buytaert et al., 2007]. There is an abundance of hydrologically disconnected areas because of the irregular topography, which gives rise to a large number of lakes and swamps [Buytaert and Beven, 2011].
Soils	Due to the steep topography, no permanent water table exists, except in local depressions where flows accumulate and permanent saturation occurs. As a result, no significant groundwater is present, and water flow is restricted to overland flow and subsurface flow in the soil layer above the bedrock [Buytaert et al., 2007]. Rainfall intensities are commonly lower than infiltration rates [Buytaert et al., 2006c]. Thus, infiltration excess overland flow (Hortonian flow) is virtually non-existent. The hydrological regime is dominated by a slow flow response [Buytaert et al., 2007]. Vertical infiltration through the soil is dominant during the beginning of rainfall events, and dependent on the antecedent soil moisture conditions. By contrast, during low intensity rainfall events, preferential flow is dominant between the organic horizon and the underlying mineral horizon or the bedrock. Saturation excess surface flow is only observed during long rainfall events; otherwise near sub-surface lateral flow in the organic layer occurs during peaks [Crespo et al., 2009]. On the other hand, surface roughness and local depressions are important in delaying surface runoff [Buytaert et al., 2006a].
Base flow	Base flow is relatively constant during the year [Buytaert et al., 2004], due to the climate, topography and soils [Buytaert et al., 2007]. Thus, the hydrological regime of the natural catchment is dominated by a slow base flow response [Buytaert et al., 2007].
Deep percolation	Subsurface groundwater is nearly absent because of the presence of impermeable bedrock [Buytaert et al., 2005b] and the lack of a groundwater storage system. Due to mountainous terrain and the impenetrable bedrock, deep percolation is negligible, and the major hydrological processes occur in the soil layer [Buytaert et al., 2004].

signatures due to different processes were represented. Finally, with the aim of analysing the impact of precipitation input uncertainty on the comparative performance of the models, these were driven by Gaussian simulated rainfall fields, and the resulting discharge ensembles were analysed using rank histograms. These were obtained through the ranking of the peak discharge of each ensemble member according to bins created with reference ensembles obtained from the other two models.

5.3.1 Modelling set up and calibration

5.3.1.1 Description of the models

Due to the availability of data in the area, the three models were run for the period 01Jul2008-31Dec2012. Data from 01Jun2008 to 01Jul2009 were used for model spin-up. In order to choose a time step for the models, the HECHMSSMA model was tested with time steps of 1 hour and 10 minutes, finding no significant differences in performance. A time step of 1 hour was used for all subsequent simulations.

A digital elevation model (DEM) of the catchments was generated from contour lines with intervals of 1, 5, 10 and 25 m (depending on the slope). The contours were processed to obtain a triangulated irregular network that was then transformed into a raster through linear interpolation. The DEM was subsequently used to delineate the sub-basins, extract morphometric parameters, and to calculate the topographic index and the channel length distribution as required by the different models.

Figure 5.2 shows the conceptual diagram of the models. A short description of each model is presented in the next paragraphs. For further details, the reader is referred to the literature cited.

TETIS is a conceptual distributed model. The estimation of runoff is based on a hydrological balance in each cell, assuming that the water is distributed into six interconnected storage tanks as shown in Figure 5.2-a. In the hills, surface flow is a combination of laminar flow and the flow occurring in a network of rills. The hydrologic processes that occur in the interrill areas and in the rills are treated jointly, in such a way that a geomorphological characterization of these elements is not needed. In parallel, interflow and base flow are generated in the corresponding soil layers. Once interflow reaches a cell with a drainage area superior to a defined threshold area for interflow, it reaches the surface, adding to the surface runoff that flows in the surface drainage network. The same occurs when the base flow reaches a cell whose drainage area is superior to the threshold for base flow. From that point on the three flows concentrate in the channel. Surface flow is then routed through the drainage network using the kinematic wave method coupled to the basin geomorphologic characteristics. The model requires the spatial estimation and calibration of the following parameters: the static storage, evapotranspiration (for this study the factor to calibrate evapotranspiration was not used), direct

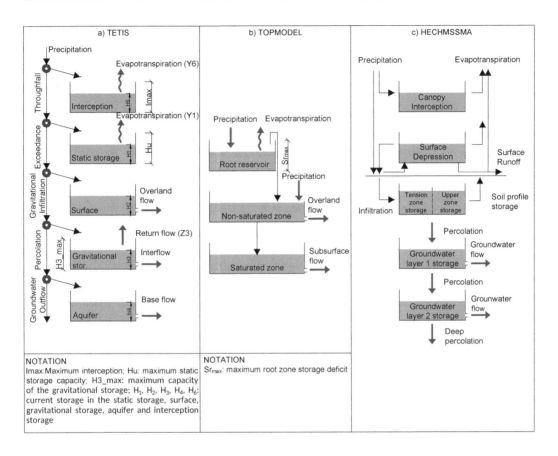

FIGURE 5.2: a) Conceptual tanks at cell level in TETIS, b) conceptual tanks TOPMODEL, c) conceptual tanks HECHMS SMA model

runoff velocity, kinematic wave velocity, infiltration rate, percolation rate, interflow velocity, base flow velocity and deep percolation rate [Frances, 2012].

TOPMODEL [Beven and Kirkby, 1979] is a semi-distributed conceptual model. Total runoff is calculated as the sum of two components (see Figure 5.2-b); saturation excess overland flow from variable contributing areas, and subsurface flow from the saturated zone of the soil [Güntner et al., 1999]. TOPMODEL uses four basic assumptions to relate down slope flow from a point to discharge at the catchment outlet: the dynamics of the saturated zone are approximated by successive steady state representations; the recharge rate to the water table is spatially homogeneous; the effective hydraulic gradient of the saturated zone is approximated by the local topographic surface gradient S ($\tan\beta$ is the notation most common in TOPMODEL descriptions, where β is the local slope angle); and the effective down slope transmissivity T of a soil profile at a point is a function of the soil moisture deficit at that point [Beven, 2012]. Flow is routed through a delay function, which represents the time spent in the channel system. The model requires the estimation of the following parameters: Initial subsurface flow per unit area, transmissivity, rate of decline of transmissivity in the soil profile, Initial root zone storage deficit, maximum root zone storage deficit, unsaturated zone time delay per unit storage deficit, and channel flow velocity inside catchment [Buytaert, 2015].

Conceptually, the HMSHMSSMA model divides the potential path of rainfall in a watershed into five tanks as shown in Figure 5.2-c [USACE, 2000]. The model simulates the movement of water through the five tanks, which represent the storage of water on vegetation, on the soil surface, in the soil profile and in the groundwater layers. Given precipitation and potential evapotranspiration (ET) the model computes basin surface runoff, groundwater flow, losses due to ET and deep percolation over the entire basin [USACE, 2000]. Twelve parameters are needed to model the hydrologic processes of interception, surface depression storage, infiltration, soil storage, percolation, and groundwater storage. The maximum depth of each storage zone, the percentage that each storage zone is filled at the beginning of a simulation, and the transfer rates, such as the maximum infiltration rate are required to simulate the movement of water through the storage zones [Fleming and Neary, 2004]. HECHMS provides several options for routing, among them the kinematic wave, which was chosen for this study.

5.3.1.2   Hydrometeorological forcing

Ordinary Kriging (OK) and Kriging with external drift were previously tested for rainfall field generation at daily scale in the study area [Rogelis and Werner, 2012]. OK was applied to a climatological variogram derived from pooled precipitation data. For Regression Kriging an average residual variogram was derived using the residuals from an average regression surface derived from a set of standardized storm data. The results of this analysis showed that the differences in performance between individual variogram interpolation with OK and with Kriging with external drift and pooled variogram interpolation are not significant. Therefore, both methods were used to obtain hourly rainfall fields for 4.5 years (July 2008 – December 2012) using the daily climatological variograms and daily average residual variograms obtained by Rogelis and Werner [2012].

The preliminary analysis of the hourly rainfall fields showed that Kriging with external drift resulted in unrealistic intensities for the study area in most storm periods (>100 mm/hr), therefore this interpolation method was not considered further. In the case of OK, runoff coefficients in the headwater catchments of the study area, showed unrealistically high values larger than 1, indicating an underestimation of the precipitation volume. In OK, when all sampling points are beyond the range of the variogram, the precipitation estimate corresponds to the mean value. Given the short ranges that characterize the convective nature of the precipitation of the study area and the sparse distribution of sub-daily rainfall gauges, most values obtained through kriging equal the mean of the recorded precipitation, leading to a significant underestimation of precipitation. However, OK rainfall fields were used as input to the models in order to identify the impact of precipitation underestimation in the models.

A second time series of rainfall fields was obtained through inverse distance weighting interpolation (IDW). The runoff coefficients obtained from these rainfall fields were in the range 0.51 to 0.56, which correspond to more realistic results.

A third time series was created in order to force the models with rainfall ensembles representing the uncertainty in precipitation inputs. This corresponds to an OK rainfall field time series, bias corrected using the IDW rainfall fields as reference time series. The bias correction was carried out through Distribution-Based Scaling - DBS [Yang et al., 2010]. This method was applied to the mean precipitation over each sub-basin to generate the bias corrected input for HECHMSSMA and TOPMODEL. In the case of the TETIS input, the bias correction was carried out pixel by pixel to obtain a bias corrected rainfall field. The bias correction procedure modifies the mean value of the rainfall field while preserving the error variance.

Conditional Gaussian simulations were obtained with the same prediction model as used for OK to create an ensemble of 50 rainfall fields, under the assumption that the variance for the original and bias corrected OK rainfall is the same. Ensembles were generated for 78 storms chosen in the period July 2009 – December 2012, which were the most significant in the basin in this period. The rainfall field ensembles were used to force the models starting from initial conditions previously estimated in a continuous simulation using the bias corrected OK rainfall fields as an input.

Hourly potential evapotranspiration fields were calculated using the Pennman FAO equation [Allen et al., 2006]. A crop factor of 0.42, as found by Buytaert et al. [2006a] in the paramos in Ecuador, was used for the areas with paramo vegetation. This was considered constant during the year, and water stress was considered non-existent [Buytaert et al., 2006c]. Daily evapotranspiration was first calculated and then a temporal distribution pattern was applied. The temporal distribution of reference evapotranspiration across the day was calculated using data of temperature, humidity, wind velocity and global solar radiation from seven hydrometeorological stations that collect data at 10 minute intervals.

5.3.1.3 Model Configuration and Calibration

Parameters were calibrated separately on a sub-basin level from upstream to down-stream in the three models. The parameters were calibrated against observed discharge measurements at the internal stations. Figure 5.3-a shows the sub-basins and the calibration points where discharge measurements are available. In order to not propagate upstream errors in the calibration process, observed discharges at upstream sub-basin outlets are used as inflow when calibrating downstream sub-basins.

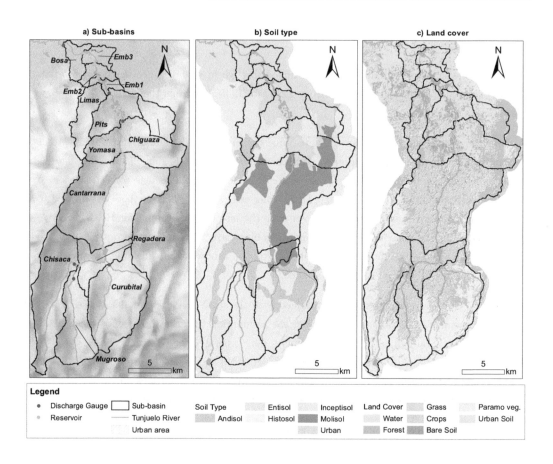

FIGURE 5.3: a) Sub-basin division; b) Soil types. Source: IGAC [2000]; c) Land cover

The initial parameters for the three models were obtained from existing soil, land cover and topo-graphical data of the basin. These are shown in Figure 5.3-b and Figure 5.3-c. Calibration was performed by optimization of the Kling and Gupta efficiency (KGE) [Gupta et al., 2009] with the Shuffled Complex Evolution (SCE) automatic search algorithm [Duan et al., 1992]. In the case of TETIS, the SCE algorithm is implemented in the software. For HECHMSSMA and TOPMODEL the SCEoptim function of the hydromad R package was used.

In the HECHMSSMA model, the Tunjuelo river basin was divided into sub-basins linked with channel reaches as shown in Figure 5.3-a. The ARCGIS HEC-GEOHMS extension [Fleming and Doan, 2013] was used for basin delineation. The initial set of sub-basins was modified to take into account the hydrological stations and the flood control structures of the river, leading to a total of 13 sub-basins with areas ranging from 4 to 92 km^2. For the watersheds in the upper basin, the hydrological stations

TABLE 5.2: 12 most sensitive HECHMSSMA calibration parameters

Parameter	Description
MaxSoilInfil	Soil maximum infiltration
MaxSoilStore	Maximum volume of the soil storage
TenStore	Tension storage
ClarkSC	Storage coefficient for the Clark's unit hydrograph
MaxSoilPerc	Maximum Soil Percolation
MaxGWStore1	Maximum Soil Percolation
MaxGWStore2	Maximum ground waters storage capacity in ground water layer 1
RoutGWStore1	Maximum ground waters storage capacity in ground water layer 2
RoutGWStore2	Groundwater flow routing coefficient in ground water layer 1
MaxPercGw1	Groundwater flow routing coefficient in ground water layer 2
MaxSoilPerc	Maximum percolation rate in ground water layer 1
MaxPercGw2	Maximum percolation rate in ground water layer 2
RoutLR12	Routing coefficient for linear reservoir 1 for baseflow
RoutLR22	Routing coefficient for linear reservoir 2 for baseflow

are located immediately upstream of the reservoirs, allowing the calibration of the entire watersheds contributing to the reservoirs.

Channel reach length and slope were determined using HEC-GEOHMS. The resolution of the DEM and absence of bathymetry prevented accurate extraction of channel cross-section information. A trapezoidal section was assumed in the middle and upper basin, with a constant Manning roughness coefficient of 0.04. In the lower basin, an average section was used according to the available bathymetry and a Manning coefficient of 0.035 was extracted from a calibrated hydrodynamic model available for the lower part of the basin.

All five tanks available in the HECHMSSMA model were used, with the Clark unit hydrograph applied as transformation method. The linear reservoir model was used for base flow estimation. With this configuration, the model has 16 parameters that require calibration in each sub-basin, as well as the initial condition of each of the five tanks. The assumption of negligible deep percolation, given the low permeability bedrock in the whole basin, reduces to 15 the number of parameters, while a warm up period eliminates the effect of initial conditions. The model parameters were first estimated based on the land cover, geology and soil information and then a three-stage calibration was carried out. First a manual calibration of the three less sensitive parameters was carried out, subsequently, SCE was used to calibrate the 12 most sensitive parameters (see Table 5.2) and finally a manual recalibration was used to refine the 3 less sensitive parameters. The sensitivity analysis of the model showed that the canopy storage and surface storage are less sensitive than the other parameters, as well as the time of concentration.

TABLE 5.3: Correction factors of the TETIS model

Correction Factor	Parameter corrected by the factor
FC_1	Static storage
FC_2	Evapotranspiration
FC_3	Hydraulic conductivity of the soil
FC_4	Surface flow velocity in the hills
FC_5	Percolation
FC_6	Horizontal saturated conductivity
FC_7	Deep percolation
FC_8	Horizontal saturated conductivity of the substrate
FC_9	Wave velocity

For areas of the basin formed by two sub-basins with only one discharge station at the outlet, distributed precipitation forcing is averaged over each sub-basin and identical model parameters are used for constituent sub-basins, thus optimizing a single parameter set.

In the case of TETIS and TOPMODEL, the effect of model resolution was explored. Parameters for the TETIS model were estimated using pixel sizes of 100, 250, 500 and 1000 meters. Smaller pixel sizes where not used due to excessive run model times. Pixel sizes of 25, 50, 100, 250, 500 and 1000 metres were used for the TOPMODEL.

Initial spatial distributed parameters for the TETIS model were estimated according to the land cover information, soils and geology [Puricelli, 2008]. Grids with the chosen resolution were created for elevation, static storage, hydraulic conductivity of the soil, percolation, horizontal saturated conductivity, horizontal saturated conductivity of the substrate, deep percolation, surface flow velocity in the hills, slope, flow direction and flow accumulation. In order to create the grids, the R project software in combination with SAGA GIS and ArcGIS was used to process the following input data: DEM of the basin; soil characteristics (sand, clay and gravel content, organic matter content, profile) according to the soil type as shown in Figure 5.3-b [IGAC, 2000]; geology [INGETEC, 2002]; and the land cover obtained from the classification of a LANDSAT Thematic Mapper 5 (TM5) image taken in 2001.

The behaviour of the water in the tanks of the model is described by equations that incorporate multiplicative correction factors for calibration purposes. The description of these correction factors is shown in Table 5.3.

TETIS uses the kinematic wave with hydraulic characteristics of the channels obtained from the geomorphological information of the watershed; this is the kinematic geomorphologic wave. An additional correction factor, FC_9 is used to correct the wave velocity. All correction factors were calibrated except for FC_2 to preserve the same input in all models. The maximum storage capacity of the gravitational tank (H3_max, see Figure 5.2-a) determines the return flow that produces saturation excess. H3_max cannot be calibrated automatically by the SCE algorithm that is hardwired in TETIS, and

TABLE 5.4: Calibration parameters of the TOPMODEL

Correction Factor	Parameter corrected by the factor
lnTe	Logarithm of the areal average of the transmissitivy
m	Model parameter controlling the rate of decline of transmissivity in the soil profile
Sr0	Initial root zone storage deficit
Srmax	Maximum root zone storage deficit
td	Unsaturated zone time delay per unit storage deficit
vr	Channel flow velocity inside catchment

a manual procedure was therefore carried out to estimate this parameter. Calibrations were carried out using maximum capacities of the gravitational storage of 10, 30, 50, 80, 100, 150 and 200% of Hu (maximum static storage capacity, see Figure 5.2-a) and a large value to completely avoid saturation excess. The sizes of the tanks are used by the TETIS model in millimetres. These variations of H3_max were tested in two model configurations: a) considering very low percolation (rock strata under the gravitational storage) therefore the aquifer tank is not used; and b) a percolation similar to the saturated conductivity, thus the aquifer tank is used in the simulations. The tests showed that for the two configurations only one of the two subsurface storages dominated the response of the watersheds. Furthermore, variations in the maximum storage do not affect the KGE coefficient and have a marginal impact on the FDC. The modifications tested in the model do not affect the overland flow, with this being minimal in all cases. The best performance of the TETIS model, from the KGE and the FDC signatures, was obtained for a model with a large capacity of the gravitational storage so no saturation excess is produced and considering a low permeability (rock strata under the gravitational storage). This was used for the subsequent phases of the analysis.

The package TOPMODEL for R [Buytaert, 2015] was used to set up the models for the three headwater sub-basins of the páramo area. The TOPMODEL application was limited to only these three watersheds, since the response of the watershed downstream is mainly dominated by the routing of the reservoir releases (see Figure 5.1), with the páramo area being the main priority for flood forecasting. A DEM with the required resolution for each sub-basin was used as input and the functions of the R package were used to obtain the topographic index distribution and the delay function. Table 5.4 shows the parameters that were calibrated.

5.3.2 Performance analysis and diagnostics

Model diagnosis is a process by which inferences are made about the representation of hydrological processes through targeted evaluation of the input-state-output response of the model [Yilmaz et al., 2008]. In order to carry out a diagnosis of the models two approaches were followed: a) an analysis

TABLE 5.5: Signature measures from the FDC [Pokhrel et al., 2012, Yilmaz et al., 2008]. Q_S and Q_O correspond to simulated and observed flows. The sub indices indicate: m1 and m2 are 0.2 and 0.7 flow exceedance probabilities; h=1,2,...H are the flow indices for flows with exceedance probabilities lower than 0.2; l=1,2,...,L is the index of the flow value located within the flow-flow segment of the FDC (0.7-1.0 flow exceedance probabilities); QS_{med} corresponds to the median value of the simulated flows and QO_{med} to the median value of the observed flows.

Signature	Description
$BiasFMS = \dfrac{[log\,(QS_{m1}) - log\,(QS_{m2})] - [log\,(QO_{m1}) - log\,(QO_{m2})]}{[log\,(QO_{m1}) - log\,(QO_{m2})]} \times 100$ (5.1)	Quantifies the % difference in the mid-segment slope of the FDC. Positive values imply that the slope of the middle portion of the simulated FDC is higher than the slope of the observed FDC.
$BiasFHV = \dfrac{\sum_{h=1}^{H} (QS_h - QO_h)}{\sum_{h=1}^{H} QO_h} \times 100$ (5.2)	Quantifies % volume bias in the highest 20% of the flows
$BiasFLV = -1 \cdot \dfrac{\sum_{l=1}^{L} [log\,(QS_l) - log\,(QS_L)] - \sum_{l=1}^{L} [log\,(QO_l) - log\,(QO_L)]}{\sum_{l=1}^{L} [log\,(QO_l) - log\,(QO)]} \times 100$ (5.3)	Quantifies the % volume bias in the lowest 30% of the flows
$BiasFMM = \dfrac{log\,(QS_{med}) - log\,(QO_{med})}{log\,(QO_{med})} \times 100$ (5.4)	Quantifies the % difference in the median flow

of the fluxes produced by each model (e.g. percolation, base flow, interflow etc) and b) the analysis of the flow duration curve (FDC) obtained from the simulated discharges at the calibration points.

Following Yilmaz et al. [2008], the flow duration curve (FDC) was used as a tool to summarize a catchment's ability to produce flow values of different magnitudes, and is therefore strongly sensitive to the vertical redistribution of soil moisture within a basin, while being relatively insensitive to the timing of hydrologic events. Five signature measures based on the FDC were used as shown in Table 5.5. The approach partitions the FDC into three segments: (1) the high flow segment, which characterizes watershed response to large precipitation events; (2) the mid-flow segment, which characterizes watershed response to moderate size precipitation events as well as the medium-term baseflow relaxation response of the watershed; and, (3) the low flow segment, which characterizes the long-term sustainability of flow [Pfannerstill et al., 2014, Yilmaz et al., 2008].

5.3.3 Analysis of precipitation input uncertainty and comparison of models

Bias corrected Gaussian simulations were used to produce a 50-member discharge ensemble for each model for the 78 chosen storms. For the models where pixel size was tested, only the best performing model resolution was used. The spread of the discharge ensembles was used as a metric of the sensitivity of the models to the variability of the precipitation. The interquartile range (IQR), the median absolute deviation averaged (MAD), and the range for all the chosen storms were calculated according to Equation 5.5, Equation 5.6 and Equation 5.7 [Franz and Hogue, 2011].

$$\overline{IQR} = \frac{1}{n} \sum_{t=1}^{n} \left(q_{0.75}\left(t\right) - q_{0.25}\left(t\right) \right) \tag{5.5}$$

$$\overline{MAD} = \frac{1}{n} \sum_{t=1}^{n} median_i \left| x_i\left(t\right) - x_{med(t)} \right| \tag{5.6}$$

$$\overline{Range} = \frac{1}{n} \sum_{t=1}^{n} \left(x_{(1)}\left(t\right) - x_z\left(t\right) \right) \tag{5.7}$$

where $q_{0.75(t)}$ and $q_{0.25(t)}$ are the 75th and 25th percentiles of the ensemble, respectively; $x_i(t)$ represents the value of a variable in each ensemble member for timestep t; $x_{med}(t)$ is the ensemble median; $x_{(1)}(t)$ and $x_{(z)}(t)$ are the lowest and highest valued ensemble members, respectively; and n is the number of timesteps.

Furthermore, rank histograms were constructed to compare the discharge ensembles between models. For each ensemble member, the peak flow was ranked using as reference the ensemble of peaks of the other two models. The peak flows of the comparison ensembles are assigned to the intervals created with the ordered peaks of the reference ensembles. Thus, the shape of the resulting histogram provides information about the ensemble in comparison with the reference ensemble. If the histogram is uniform the two ensembles are similar, if the histogram is skewed to the right the comparison ensemble tends to higher values than the reference ensemble and the opposite if it is skewed to the left.

Once the frequency of the peak discharges of the comparison ensemble has been determined according to bins created with the reference ensembles, all the rank histograms are pooled obtaining the frequency of the ensemble peaks of each model according to the ordered ensemble peaks of the other two models.

5.4 Results

5.4.1 Model calibration

5.4.1.1 KGE for HECHMSSMA, TOPMODEL and TETIS

The first two columns for each sub-basin in Table 5.6 show the optimum KGE values obtained from calibration using as forcing for the models the OK and IDW rainfall fields. The third column (OKbc) shows the KGE obtained from the simulations with models configured with the parameters obtained from calibration with IDW rainfall fields but using the bias corrected OK rainfall fields as input precipitation. In the case of the TOPMODEL, the KGE for OKbc was obtained for grid sizes of 500 m and smaller, due to the drop in performance for larger grid sizes.

There is an increase in performance when using IDW rainfall fields in comparison with OK rainfall fields for the Mugroso and Curubital sub-basins. For the Chisacá sub-basin the increase in performance occurs for the HECHMSSMA model and for the TETIS model with pixel sizes smaller than 500 m. In the case of the sub-basins located in the middle and lower basin the differences are less significant. The use of OKbc as forcing for the models produces minor reductions of efficiencies when compared with the best efficiency obtained for IDW precipitation and a pixel size of 500 m in the case of HECHMSSMA and TETIS; and very similar values in the case of the TOPMODEL.

The calibration results can be grouped into poor performance ($0.5 > KGE > 0$), intermediate ($0.75 > KGE > 0.5$) and good performance ($KGE > 0.75$) [Thiemig, 2014]. According to this classification the headwater catchments located in the paramo area (Chisaca, Mugroso and Curubital sub-basins), exhibit maximum efficiency values in the range of intermediate performance (see Table 5.6). The maximum efficiency values are similar for all the three models, with HECHMSSMA and TETIS reaching the highest values compared to TOPMODEL. Regarding the watersheds downstream of the páramo area, the results are dominated by the discharge from the reservoirs, and therefore depend mainly on the routing of the measured hydrograph. KGE values are in the range of intermediate to good performance.

5.4.2 Comparison of water balance fluxes

The total volumes of the fluxes in millimetres from each model and for IDW and OK rainfall fields are shown in Figure 5.4. The results obtained by driving the models with OKbc rainfall fields are not shown since they are similar to the results obtained from IDW rainfall fields. Only the results for the headwater watersheds in the páramo area are shown, since the release of the reservoirs dominates the outflow discharge of the watersheds downstream.

Model	Cantarrana			Chisaca			Curubital			Independencia			Mugroso			SnBenito		
	OK	IDW	OKbc	OK	IDW	OKbc	OK	IDW	OKbc	OK	IDW	OKbc	OK	IDW	OKbc	OK	IDW	OKbc
TETIS																		
100	0.85	0.81		0.57	0.60		0.44	0.63		0.84	0.84		0.43	0.64		0.64	0.75	
250	0.85	0.80		0.56	0.59		0.45	0.64		0.84	0.83		0.46	0.64		0.64	0.70	
500	0.85	0.81	0.78	0.59	0.58	0.57	0.45	0.65	0.59	0.84	0.84	0.79	0.51	0.67	0.66	0.67	0.75	0.75
1000	0.84	0.80		0.64	-0.24		0.25	0.59		0.84	0.84		0.51	0.67		0.79	0.84	
2000	0.85	0.78		0.62	-0.22		0.41	0.63		0.85	0.84		0.50	0.69		0.90	0.90	
Topmodel																		
25				0.54	0.57	0.59	0.43	0.62	0.59				0.46	0.63	0.65			
50				0.57	0.57	0.58	0.43	0.62	0.59				0.45	0.62	0.65			
100				0.58	0.57	0.58	0.43	0.61	0.58				0.45	0.62	0.65			
250				0.58	0.57	0.57	0.42	0.61	0.58				0.45	0.62	0.64			
500				0.57	0.55	0.56	0.41	0.61	0.58				0.44	0.60	0.63			
1000				0.51	0.50		0.42	0.61					0.42	0.58				
HEC-HMS																		
SMA	0.78	0.74	0.73	0.44	0.58	0.54	0.34	0.65	0.56	0.88	0.91	0.91	0.38	0.65	0.6	0.60	0.67	0.66

TABLE 5.6: Kling and Gupta coefficient obtained from calibration

Results show that the input rainfall obtained through OK is lower in comparison with the precipitation obtained from IDW, leading to models where there is no actual evapotranspiration, which highlights the underestimation of precipitation values by the OK interpolation. This underestimation significantly affects the performance of the models, not only in terms of efficiency as shown in Table 5.6 but also affects the models' ability to properly simulate hydrological processes.

The results of the HECHMSSMA model show a dominance of the groundwater flow from the groundwater layer 2 for Chisaca and Mugroso and overland flow for Curubital for OK rainfall fields and dominance of the groundwater flow from the groundwater layer 2 for Mugroso and Curubital and overland flow for the case of Chisaca when IDW rainfall fields are used.

In the case of TOPMODEL, the response of all the models is dominated by subsurface flow with a small contribution of overland flow. When OK rainfall fields are used the response in the three watersheds does not vary significantly and the evapotranspiration is negligible. When the precipitation is calculated with IDW, the evapotranspiration is significant in the water balance with approximately equal proportions in the Mugroso and Curubital sub-basins and approximately twice in the Chisaca sub-basin.

The TETIS model has a similar behaviour for both the IDW rainfall fields and the OK rainfall fields with the latter being lower in volumes. The dominant process in the headwater páramo catchments is interflow, this is the release from tank 4 (gravitational storage) in Figure 5.2-a. In the case of the IDW rainfall fields overland flow is negligible. In contrast, this flux is observable when OK fields are used, albeit in a small volume. The increase in pixel size influences the proportion of subsurface flow and evaporation. The most severe changes are observed for a pixel size of 1000 m. These are due to a significant change in the drainage area and stream network caused by the coarse grid. In the case of the Chisaca sub-basin, for pixel sizes higher than 500 m, the drainage area duplicates increasing the precipitation input, which causes a significant increase in the storage when IDW rainfall fields are used and an increase in evapotranspiration in the case of OK rainfall fields to compensate for the

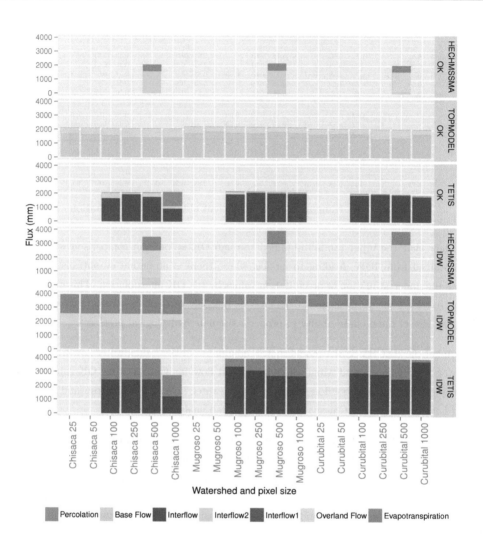

FIGURE 5.4: Fluxes obtained from the models

increase in drainage area, due to changes in connectivity. This resolution is, thus, considered to be too coarse for these small watersheds.

Considering only pixel sizes up to 500 m, when input precipitation obtained from IDW is used, in the case of Mugroso and Curubital watersheds, actual evapotranspiration reduces with pixel size, while in the case of Chisaca the evapotranspiration accumulation remains constant.

5.4.3 Signature measures from the flow duration curve (FDC)

Due to the significant underestimation of precipitation with the OK interpolation, the results in this section will refer only to the precipitation obtained through IDW interpolation.

Figure 5.5 shows the flow duration curves from the HECHMSSMA, the TOPMODEL and TETIS the models. Only the results for the pixel size of 25 m are shown for TOPMODEL since the curves for the other sizes are similar. For the case of TETIS only the models up to a pixel size of 500 m were considered due to the significant deterioration of the representation of the drainage network occurring when larger pixel sizes are used.

In the case of TOPMODEL the lowest overall biases for all the watersheds are found for a pixel size of 25 meters, as well as the highest KGE. This pixel size was therefore used for subsequent analysis.

The TETIS model better represents the high flow portion of the duration curve (discharges equalled or exceeded less than 20 % of the time) exhibiting the lowest bias values (FHV) in the case of the Mugroso and Curubital watersheds (see Table 5.7). For the Chisaca watershed, the TOPMODEL has a significantly better performance than the other models in this part of the FDC exhibiting the lowest bias values.

The middle portion of the FDC (flows equalled or exceeded between 20 and 70 % of the time, see vertical lines in Figure 5.5) is better represented (lowest FMS, see Table 5.7) by the TOPMODEL in the Chisaca and Curubital and by the TETIS model in the case of the Mugroso watershed.

The TETIS model exhibits the highest biases (FLV) for the lowest flows (flows equalled or exceeded more than 70 % of the time), while the lowest biases correspond to the TOPMODEL.

In terms of grid size, the lowest overall biases in the TETIS model are obtained for a grid size of 100 m for Mugroso and Curubital and for a grid size of 500 m for the Chisaca sub-basins (see Table 5.7). However, the lowest biases in the high flow segment of the FDC for Mugroso and Curubital correspond to the grid size of 500 m. A grid size of 500 m was chosen for the subsequent analysis due to its good representation of high flows, and insignificant reduction of performance in the middle and low segment of the FDC. Models until this grid size also had a shorter computation time and higher KGE values.

5.4.4 Rainfall ensemble analysis, input precipitation uncertainty

The results in Table 5.6 , show that the bias corrected OK rainfall fields provide a very similar response of the models to that found with the IDW rainfall fields. Given the good performance of the bias correction, the Gaussian simulations were produced by applying the climatological variogram used in

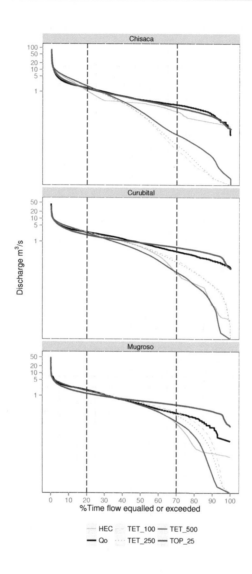

FIGURE 5.5: Flow duration curves and signature measures

the OK interpolation, and then bias corrected using the corresponding mapping functions, and used as input to the models to test their sensitivity to variability in precipitation.

Table 5.8 shows the IQR, MAD and range for the ensemble discharge of the 78 storms selected in the period of analysis. In all three watersheds, the metrics calculated for all storms have similar values in each watershed, with the highest values consistently corresponding to the TOPMODEL, except for the MAD and IQR for the Chisaca sub-basin, where the highest values correspond to the HECHMSSMA

Model	FHV	FLV	FMS	FMM
Chisaca				
TETIS 100	31.62	274.11	270.92	181.95
TETIS 250	32.68	261.76	245.22	175.84
TETIS 500	34.96	186.09	188.52	144.07
HECHMSSMA	22.28	34.20	24.33	41.95
TOPMODEL 25	5.06	10.70	12.72	6.24
TOPMODEL 50	4.01	11.31	13.45	6.89
TOPMODEL 100	3.65	11.51	14.39	9.37
TOPMODEL 250	2.61	11.32	13.34	9.77
TOPMODEL 500	− 1.68	17.93	17.90	11.06
Mugroso				
TETIS 100	− 2.05	54.77	− 4.83	−19.12
TETIS 250	− 2.54	75.47	7.33	− 4.85
TETIS 500	− 1.77	113.77	33.25	17.80
HECHMSSMA	− 6.43	76.04	41.61	4.98
TOPMODEL 25	−20.16	−47.42	−50.10	−35.34
TOPMODEL 50	−20.57	−48.36	−50.66	−37.36
TOPMODEL 100	−20.57	−47.90	−50.50	−36.12
TOPMODEL 250	−20.72	−48.60	−50.80	−37.78
TOPMODEL 500	−21.57	−49.92	−52.02	−41.47
Curubital				
TETIS 100	4.98	127.12	45.72	207.02
TETIS 250	5.71	135.51	56.64	244.07
TETIS 500	2.67	247.47	103.26	533.48
HECHMASMA	5.87	213.19	118.50	245.71
TOPMODEL 25	−10.33	−27.66	−34.88	− 5.34
TOPMODEL 50	− 9.91	−30.33	−36.54	−34.56
TOPMODEL 100	−10.05	−32.26	−37.53	−46.09
TOPMODEL 250	−10.10	−32.60	−37.65	−54.08
TOPMODEL 500	− 9.93	−32.30	−37.93	−51.87

TABLE 5.7: Flow duration curve signature measures

model. The smallest ensemble spreads are found in the Mugroso sub-basin, while the highest are found in the Chisaca sub-basin. The average values for the peak discharge of each storm shows that the TOPMODEL is clearly the most sensitive model to variations in rainfall input, exhibiting the largest IQR, MAD and Range at the peak.

5.4.5   Comparison of model ensembles

Figure 5.6 shows the rank histograms for the head watersheds in the páramo area comparing the discharge ensembles of the models. The comparison of the ensembles obtained from HECHMSSMA

TABLE 5.8: IQR, MAD and range of ensemble discharges for HECHMSSMA, TOPMODEL and TETIS

	IQR	MAD	Range	IQR_{peak}	MAD_{peak}	$Range_{peak}$
Curubital HMS	3.70	1.89	13.45	6.36	3.31	22.94
Curubital TET	3.58	1.78	14.75	7.19	3.60	28.27
Curubital TOP	4.16	2.00	16.74	15.03	7.27	68.47
Mugroso HMS	1.39	0.66	5.75	2.20	1.06	9.11
Mugroso TET	1.62	0.69	7.06	2.78	1.24	11.87
Mugroso TOP	1.77	0.79	8.62	6.17	2.74	36.59
Chisaca HMS	4.19	2.17	17.56	10.02	5.50	40.94
Chisaca TET	3.27	1.72	16.66	8.16	4.29	40.75
Chisaca TOP	3.80	1.85	17.86	16.00	8.06	76.31

and TOPMODEL (first column in Figure 5.6) shows that the members of the TOPMODEL ensemble have mostly higher values than the HECHMSSMA ensemble. The comparison of TETIS and TOP-MODEL shows equally that the members of the TOPMODEL have mostly higher values than the TETIS ensemble. In the case of TETIS and HECHMSSMA, the rank histogram shows less difference between the two ensembles with an approximately uniform distribution for the Mugroso and Curubital watersheds. For the Chisaca watershed, the rank histogram shows underdispersion meaning that most values of the HECHMSSMA model are larger or smaller than the TETIS ensemble.

5.5 Discussion

5.5.1 Model calibration and performance

5.5.1.1 Water balance fluxes and hydrometeorological forcing

The precipitation and evapotranspiration data are considered as the main source of uncertainty in the models (Buytaer et al. 2005). Precipitation data in the páramo area are subject to errors inherent to the significant difficulties in the measurement process and high spatial rainfall variability [Buytaert et al., 2006c]. Wind speeds at high altitude may be high and a smaller or larger portion of the rain may be blown over the rain gauge [Sevink, 2007]. Furthermore, fog is highly difficult to quantify [Bruijnzeel, 2001, Tobón and Gil - Morales, 2007], and this may add an unknown quantity of water, especially where patches of arbustive species are present [Buytaert et al., 2006c].

Evapotranspiration is influenced by the particularly low evaporation characteristics of the vegetation. Tobón and Gil - Morales [2007] found that during those fog events that do not produce dripping onto the floor, there are no net inputs to the ecosystems, and the contribution of fog to the catchment water yield can be only through their control over forest transpiration.

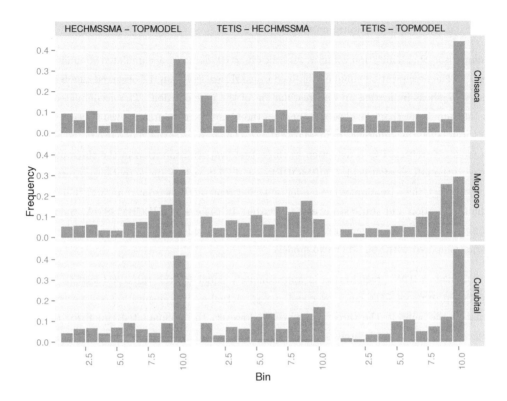

FIGURE 5.6: Rank histograms for the head watersheds in the páramo area for the three models. The bins were reduced to 10 for better visualization. The first model in the title corresponds to the reference ensemble, e.g. in HECHMSSMA - TOPMODEL the reference ensemble corresponds to the results of the HECHMSSMA model and the ensemble to be compared is the obtained from TOPMODEL simulations.

Even though the approximation of evapotranspiration by using the Penman FAO equation is considered appropriate for paramo areas by some authors [Sevink, 2007], difficulties in the reliable estimation of humidity under foggy conditions [Sevink, 2007] may introduce significant errors. In addition, fog is not only thought to induce an extra input of water into the ecosystem but also to suppress evaporation [Sevink, 2007]. Buytaert et al. [2006a] highlights the limited validity of the Penman FAO equation under the unusual meteorological conditions of the páramo.

Besides the impossibility to include fog interception given the lack of data, the estimation of the rainfall field has shown to be highly challenging. Different interpolation methods lead to significantly different precipitation volumes, strongly influencing the efficiency and performance of the models. Ordinary kriging using a daily climatological variogram produces lower KGE values than IDW; this is mainly due to the underestimation of precipitation volumes in the case of the former. IDW seems to produce more realistic precipitation values.

The comparison of precipitation volumes with the observed discharge accumulated volumes is shown in Figure 5.7. The precipitation volumes obtained from OK are less than the observed runoff in the three paramo watersheds indicating an underestimation of the precipitation. The accumulated rainfall is about 1000 mm lower than the observed runoff in the Mugroso and Curubital watersheds, while in the Chisaca watershed this difference reduces to 200 mm. The impact of this difference in the performance of the models is reflected in a reduction in the actual evapotranspiration in the three models when forced with the OK rainfall field. TOPMODEL and TETIS reduce the actual evapotranspiration to almost 0 through the reduction of the model tank that represents the root zone storage in the former, and the interception and static storage in the latter. In the case of HECHMSSMA, the model does not completely reduce the evapotranspiration, which then leads to a significant underestimation of discharge compared with the other two models.

The accumulated IDW precipitation is approximately equal for the three watersheds. Given the similarities in terms of soils, land cover and geology actual evapotranspiration is expected to be approximately the same in the three watersheds. However, the accumulated runoff for the Chisaca watershed is lower (approximately 1000 mm lower than in the other watersheds) which leads to a resulting actual evapotranspiration that is higher than in the other watersheds. This suggests a relative overestimation of precipitation (real precipitation lower than the precipitation in Mugroso and Curubital watersheds) for the Chisaca watershed that produces an increase in the actual evapotranspiration to balance outputs in the models. This behaviour of the Chisaca watershed suggests that the available precipitation data is not representative of the precipitation occurring in this watershed.

According to Buytaert et al. [2006a] literature values of calculated actual evapotranspiration for grass páramo range from 0.8 to about 1.5mm/day. The only two models in or close to that observed range are the TOPMODEL and TETIS forced with IDW rainfall fields with values of 0.82 and 0.89 mm/day respectively for Chisaca, 0.78 and 0.73 for Mugroso and 0.5 and 0.86 for Curubital. In the case of the other models the actual evapotranspiration is highly underestimated in comparison with observed values reported in literature. These results show that realistic ranges of actual evapotranspiration are only obtained in the Chisaca watershed and in the Curubital watershed with TETIS, suggesting that the precipitation volume estimated with IDW is low mainly for Curubital and Mugroso.

5.5.1.2 Pixel size and flux variation for the TOPMODEL and TETIS

In the following sections the discussion will focus only on the results obtained from the calibration using IDW rainfall fields, due to the underestimation obtained when OK rainfall fields are used.

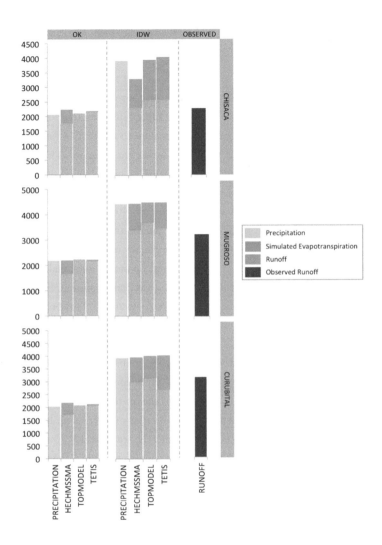

FIGURE 5.7: Water balances Chisaca, Curubital and Mugroso watersheds

TOPMODEL and TETIS models with pixel sizes larger than 500 m produce similar, and in some cases better KGE than in the case of finer grids. However, the drainage network of the watersheds cannot be correctly represented with these pixel sizes in the watersheds in the paramo areas. The most notorious case corresponds to the Chisaca and Mugroso watersheds where the distance between the two main streams is less than 1000 m in some reaches, which leads to accumulation grids that cannot correctly represent the stream network. The similar KGE are due to adjustment of the model parameters without correctly representing the hydrological behaviour of the watersheds. The results of these coarse models will also not be taken into account further.

In TOPMODEL the KGE values vary in a maximum range of only +/- 0.03, reflecting very similar efficiencies regardless of the pixel size. Increasing the grid size of the DEM increases mean values of the topographic index [Deginet, 2008]. The mean topographic indices for the three watersheds increase when the grid size increases going from values close to 6.5 to 10.4 for the coarser resolution (1000 m), increasing more significantly for pixel sizes larger than 100 m. This is due to the greater upslope contributing area and smaller slope [Wu et al., 2007]. This behaviour in the topographic index is consistent with previous studies [Bruneau et al., 1995, Deginet, 2008]. Wu et al. [2007] found that the smoothing effect of grid size increase may result in deteriorated topographic index distributions at coarse resolutions. However, this can be moderated by parameter calibration, as found in the results shown in Table 5.6. Despite the change in the topographic index distributions, fairly similar efficiencies can be preserved by the compensation effect of the calibration parameters, mainly transmissivity. The increase in grid size produces an increase in saturated areas that results in the increase of overland flow when the same calibration parameters are kept [Deginet, 2008]. This behaviour is explained by the disappearance of the smaller values and increase of the mean values of the topographic index [Deginet, 2008]. Hence, the adjustment of the transmissivity to higher values allows to obtain almost identical model efficiencies [Franchini et al., 1996, Saulnier et al., 1997, Wu et al., 2007]. The increase in transmissivity is larger for pixel sizes larger than 100 m, in correspondence with the increase in the topographic index. This increase in transmissivity keeps the overland flow proportion fairly similar for the three watersheds. The calibrated transmissivity values for pixes sizes up to 500 m range between 0.32 and 16.5 m2/h. The lowest values are consistent with the transmissitivy values found by Buytaert et al. [2005a] for páramos in Ecuador and the highest values are still in the range of transmissivity values found in other applications of the TOPMODEL [Beven, 1997].

In the TETIS model the variation of pixel size produces only minor changes of +/-0.03 in the KGE. However, a pixel size of 500 m is an optimum in terms of KGE in the case of Mugroso and Curubital. In all three watersheds, the lateral conductivity of the soil increases with pixel size. This is the main parameter used by the model to compensate for variations in pixel size. The discharge coefficient (α), that multiplies the storage in tank 3 (H3) to obtain its outflow (interflow) (see Figure 5.2-a), is a function of the horizontal saturated conductivity, the pixel size and the time interval. This mathematical relation explains its scale dependency. The values of the discharge coefficient in the paramo watersheds range from 0.72 to 0.89 implying high outflows from tank 3.

The comparison of the calibration results of the three watersheds in terms of the behaviour of each tank (see Figure 5.2) can be summarized as follows:

a) Static storage: this storage corresponds to water that can be evaporated from surface depressions, vegetation and water retained in the soil through capillary forces [Frances, 2012]. The correction factors that multiply the capacity of the storage correspond to values higher than 1 for Chisacá, but remain approximately constant for pixel sizes from 100 to 500. In the case of Mugroso, correction factors increase with pixel size but remain low, reaching 0.12 for a pixel size of 500 m. For Curubital correction factors slowly increase with pixel size to reach a value close to 1 for a pixel size of 500 m. Due to the close connection of this storage with the evapotranspiration process, these results may be due to a lack of representativity of the precipitation data obtained from the station located most to the south-west of the study area (see Figure 5.1), that may lead to relative underestimation of precipitation in the Chisaca watershed, forcing the model to compensate by increasing the capacity of the tank to increase evapotranspiration losses in comparison with the other watersheds (see Figure 5.4). b) Superficial storage: The calibrated hydraulic conductivities are high in comparison with the rainfall intensities of the páramo area. For a pixel of 500 m calibrated hydraulic conductivities range between 82-135 mm/h, which correspond to high values in comparison with a range of 10-60 mm/h found in other páramo areas [Crespo et al., 2009]. These high conductivity values result in no infiltration excess occurring in the model, and the water moving to the gravitational storage, which is consistent with the characteristics of the páramo described in Table 5.1. c) Gravitational storage: the calibrated percolation is very low, which means almost no water is going to the aquifer storage. This explains why the flow is dominated by the outflow from this tank (see Figure 5.4). The fluxes of the model are dominated by the discharge from the tank number 3, which can be interpreted as the discharge from shallow soil above impervious strata. No saturation excess flow is produced; therefore the model does not simulate any rapid response/overland flow of the watershed. The behaviour of the gravitational storage is coherent with the hydrological behaviour of páramo watersheds described in Table 5.1. d) Aquifer storage: due to the very low permeability of the rock underneath the soil layer, the storage in this tank is negligible, as well as the outflow.

5.5.1.3   HECHMSSMA calibration results and fluxes

The KGE values obtained from the HECHMSSMA are similar to the values obtained from the calibration of the TETIS model. In terms of fluxes of the models (see Figure 5.4), these are similar for the Mugroso and Curubital watersheds for IDW rainfall fields (dominance of subsurface flow). Conversely, the response of the Chisacá watershed is dominated by overland flow generated through infiltration excess. This is due to a low soil infiltration in the calibrated model, producing a response dominated by overland flow. This representation of processes in the model is not consistent with the other two models, or with the perceptual model of the watershed, implying the inability of the HECHMSSMA model to adequately represent the hydrology of the Chisacá watershed, given the available data. From the previous analysis, a relative overestimation of precipitation was detected in the Chisaca watershed, which suggested a lack of representativity of the measured precipitation in this watershed. This difference in hydrometeorological forcing may be the cause for different hydrological processes calibrated to represent the watershed response.

The response of the Mugroso and Curubital rivers is similar. Saturated conductivities larger than the rainfall intensity prevent the occurrence of overland flow. The soil percolation rates are high, therefore infiltrated water moves rapidly to the first ground water layer. The percolation rate from the first groundwater layer to the second is high, therefore water moves quickly to the second groundwater layer. This rapid percolation to the second groundwater layer inhibits outflow from the first groundwater layer; therefore the subsurface response is dominated by the outflow from the second groundwater layer. This behaviour is consistent with the dominance of subsurface flow, characteristic of páramo areas described in Table 5.1.

5.5.1.4 Flow duration curve and signatures

The sensitivity to variations in pixel size is negligible in the case of TOPMODEL where the transmissivity parameter compensates changes in grid size, reaching similar KGE values and producing very similar FDC with almost the same signatures, with the finer pixel model showing the smallest biases. Conversely, the TETIS models are significantly sensitive to changes in pixel size mainly, for low discharges (equalled or exceeded more than 70% of the time). Furthermore, the TETIS model exhibits the poorest performance for low discharges. This is due to the rapid outflow from the storage representing the subsurface flow, which fails to represent the slow water release of the soil of the paramo areas. The same behaviour is observed in the HECHMSSMA model for the Mugroso and Curubital sub-basins. However, in the Chisaca sub-basin that model better represents low discharges when compared to TETIS, since the response of the model is dominated by infiltration excess, and the subsurface flow is modelled through the water release from the second underground storage with a large routing coefficient. In the TETIS models and in the HECHMSSMA models, the subsurface flow is represented by only one storage, despite having the possibility to use two. In both models, the water flows rapidly to the deeper storage that controls the response.

In general, the TETIS models overestimate discharges for large discharges (equalled or exceeded less than 20% of the time) and underestimates for lower discharges. The HECHMSSMA model has a similar behaviour. The TOPMODEL overestimates low discharges in the case of Mugroso and Curubital and slightly underestimates them in the case of the Chisaca watertheshed. For high discharges, TOPMODEL has a good representation of the FDC in the case of Chisaca and Curubital sub-basin, and a slight underestimation in the case of Mugroso.

Given the FDC results and the KGE, TOPMODEL appears to be the most realistic model of the three models tested in this analysis. This is supported by the assumptions of TOPMODEL that seem to be able to adequately represent the main characteristics of the paramo soils response [Buytaert and

Beven, 2011], with the hydrologic response dominated by the topography and no infiltration excess overland flow; and nonlinear transmissivity profile. In agreement with other studies carried out in the páramo area [Buytaert and Beven, 2011] the assumption of an exponential function of the storage deficit seems to provide a good representation of the processes in these watersheds.

According to the results, higher performance metrics such as the KGE do not necessarily mean a better representation of hydrological processes, and therefore, they are not an indication of realism of the model, which is necessary for flood forecasting reliability [Kavetski and Fenicia, 2011]. However, the use of signatures and analysis of model fluxes provides a means to compare model structures in terms of their abilities and limitations to reproduce the dominant hydrological processes, and to gain insight into the characteristics of a model that make it more suitable than others. Consistency, defined by Euser et al. [2013] as the ability of a model to reproduce several hydrological signatures with the same parameter set is a criterion that provides the means to assess the reliability. Furthermore, the comparison in terms of process representation is crucial to interpret the effects of using different model structures [McMillan et al., 2011]. The correspondence between catchment structure and model structure was identifiable in this study, which provides understanding about the watersheds behaviour.

5.5.2 Comparison of discharge ensembles

The analysis of the discharge ensemble spread in the models shows a higher sensitivity of TOP-MODEL to variation in the rainfall. Increases in the precipitation cause a significant increase of the peak discharges of the storms since the precipitation over saturated areas immediately contributes to overland flow. For the TETIS and HECHMSSMA models precipitation infiltrates and flows as subsurface flow through the underground tanks, which reduces the increase in peak flow in comparison to TOPMODEL. This means that, TOPMODEL is the most sensitive model to rainfall variability, albeit the most realistic.

The Chisaca sub-basin shows the larger ensemble spread metrics, with this being the sub-basin with the most unreliable precipitation input. Due to the apparent relative overestimation of precipitation in this watershed, the parameters of all models adjust to increase evapotranspiration and reduce the outflow discharge. Therefore when increases in precipitation occur in the rainfall ensembles, the increases in peak flows are larger than in the other models, where the balance between the fluxes seems more realistic.

5.6 Conclusions

A distributed model (TETIS), a semi-distributed model (TOPMODEL) and a lumped model (HEC HMS soil moisture accounting) were used to simulate the discharges of a tropical high mountain basin characterized by soils with high water storage capacity and high conductivity. The performance analysis and diagnostic applied allowed insight in the representatively and appropriateness of the models. The comparison of models, through performance measures combined with analysis of fluxes and flow duration curve signatures, provided a means to assess the abilities and limitations of the models. This analysis allows insight into the models process representation, providing the information needed to identify a model structure that is more suitable than the others in terms of how realistically relevant hydrological processes are simulated.

Different model structures were shown to have similar performance according to the King and Gupta efficiency (KGE) value, however their ability to reproduce hydrological processes varies. The ability to reproduce hydrological processes is also influenced by inputs errors. Overestimation and underestimation of precipitation can produce a change in the dominant hydrological processes simulated by the models, with some models more sensitive to these errors than others. In the study area, the use of a climatological variogram with ordinary kriging to interpolate hourly rainfall fields proved to result in underestimation of rainfall, significantly affecting the performance of the models. Due to the complex spatio-temporal variability of precipitation, the simpler approach, using Inverse Distance Weighting (IDW) was found to be the most appropriate.

The use of varying pixel sizes in the semi-distributed and distributed model, showed that a first and determinant criteria for upper limits in pixel size is the ability of the grid to appropriately reproduce the drainage characteristics of a basin. Furthermore, variations in the pixel size are compensated by selected parameters in each model, in order to reach approximately the same performance for all grid sizes. In the case of TOPMODEL the compensation is achieved though variations in the transmisitivy, for TETIS the compensation is manly achieved through variations in the lateral conductivity of the soil.

Despite the compensation of parameters, an optimum grid size could be identified in the TETIS and TOPMODEL through the use of the FDC signatures, through which the slight variations in representation of processes could be identified according to pixel size. These optimum grid sizes are 500 meters for TETIS and 25 meters for TOPMODEL.

The behaviour of TETIS and HECHMSSMA models for the páramo is similar in terms of the water flow in the underground tanks. Only one of the two underground tanks available is used due to high conductivity values that produce a rapid flow towards the deeper tank. In the case of TETIS the tests with several configurations of the model showed that a model consisting of a tank representing the soil layer over an impervious rock layer (aquifer storage) performs best. This is consistent with the perceptual model of the hydrology of the watershed. In the case of HECHMSSMA one of the two tanks representing interflow dominates the response. However, saturation excess is not modelled by any of these two models, thus the flow is exclusively dominated by the release of one underground tank. With this configuration, none of these models has the ability to reproduce the slow water release in the low flow portion of the FDC. This is due to a rapid flow of water from the dominating underground tank in response to the high conductivities that are obtained from calibration. For these models, even if a relatively good representation of high discharges can be achieved, low flows cannot be modelled appropriately.

TOPMODEL appears to be the most realistic model for the páramo of the models tested in this analysis, although it is more sensitive to rainfall fields variability. This model is able to reproduce the slow water release from the soil layer over the rock stratum that is one of the main characteristics of the páramo soil. The signatures obtained from the flow duration curves show that this is the model that more closely reproduces all ranges of discharge in the three páramo sub-basins. Besides providing more reliability, TOPMODEL demands low computational resources and short run times. These aspects support that TOPMODEL is the preferred choice from a flood early warning perspective.

Chapter 6

Streamflow forecasts from WRF precipitation for flood early warning in tropical mountain areas

This chapter is an edited version of: Rogelis, M. C., and Werner, M. G. F. : Streamflow forecasts from WRF precipitation for flood early warning in tropical mountain areas, to be submitted to Hydrological Processes.

6.1 Introduction

Numerical Weather Prediction (NWP) models are fundamental to extend lead-times beyond the concentration time of a watershed. The significant advances in NWP and computer power during the last decades have led to the generation of high resolution precipitation forecasts at the catchment scale. Therefore, quantitative precipitation forecasts (QPF) from high-resolution NWPs are increasingly used in flood forecasting systems [Cluckie et al., 2006]. Particularly, flash flood forecasting systems typically needed in tropical mountainous watersheds require forecast precipitation to provide timely warnings.

Despite the significant advances, NWP results contain noise, are contaminated by model biases, are too coarse to adequately resolve all features such as convection, and are influenced by uncertainty inherent in the initial conditions [Colman et al., 2013]. Furthermore, weather forecasting in tropical mountains is highly challenging. In the tropics, local and mesoscale effects are more dominant than synoptic influences (except for tropical cyclones). In addition, there are limitations in the availability of surface and upper air monitoring networks [Laing and Evans, 2010], which influences modelling

134

initialization [Cuo et al., 2011]. Regarding orography, this influences the formation and movement of deep convection and mesoscale convective systems [Colman et al., 2013].

According to Habets et al. [2004] the potential of NWP precipitation forecast to be used by hydrological models for flood forecasting is mainly affected by: (i) localisation of the events, since an error of a few kilometers can lead the precipitation in the wrong watershed; (ii) timing of the events, since the response of the basin depends on previous events and on the timing of the present event; and (iii) precipitation intensity. This is especially true in flash flood prone watersheds, typical of tropical mountainous areas where location errors that are considered small at meteorological level can lead to completely miss a flood event [Vincendon et al., 2011].

In order to address the uncertain nature of NWP, ensemble prediction systems (EPS) have been developed [Demeritt et al., 2007]. In contrast to deterministic systems that produce one prediction, EPS produce a suite or ensemble of predictions to reflect uncertainty, providing the capability to transform predictions into a probability distribution function [Demeritt et al., 2007, Leutbecher and Palmer, 2008]. As EPS rainfall predictions often exhibit greater skill than deterministic predictions, the hope is then that EPS products will increase the skill and time horizon of flood forecasts [Demeritt et al., 2010].

The use of ensembles of NWP models to drive flood forecasting systems has increased and is a relevant research topic [Cloke and Pappenberger, 2009]. Currently many flood forecasting centres use ensemble prediction systems (EPS) for representing uncertainty, but most of these are in Europe, Canada, The United States and Australia [Demeritt et al., 2007]. Experience of this kind of system in developing countries is very limited [Fan et al., 2014].

Rainfall forecasts provided by NWP require post-processing to correct bias and to reliably quantify uncertainty [Robertson et al., 2013]. Several approaches have been used to produce ensemble rainfall forecasts by post-processing raw numerical weather prediction (NWP). Robertson et al. [2013] present a method that uses a simplified version of the Bayesian joint probability modelling approach to produce forecast probability distributions for individual locations and forecast lead times; Theis et al. [2005] propose a methodology based on the hypothesis that some probabilistic information about a precipitation forecast at a certain time and location can be derived from its spatio-temporal neighbourhood in the model precipitation field. A set of forecasts is extracted from the spatio-temporal neighbourhood of a point and used to derive a probabilistic forecast at the central point of the neighbourhood; Bremnes [2004] proposes a method to produce probabilistic forecasts in terms of quantiles from NWP output using probit regression and quantile regression; Clark et al. [2004] use a two-stage approach that includes logistic regression and ordinary least squares regression to generate precipitation and temperature ensembles. Probabilistic forecasts of precipitation are especially challenging since precipitation has a mixed discrete-continuous probability distribution [Frei, 2012]. In this study, a simple two-stage approach is used to produce precipitation ensembles from WRF forecasts, based on probit regression and quantile regression.

The assessment of the value of forecasts is a crucial issue. Verification is essential for the understanding of the abilities, weakness and value of forecasts, which leads to improving the forecast system. Several scores are proposed in literature to asses the quality of forecasts as well as skill scores aimed at quantifying the relative accuracy of a forecast with respect to a reference forecast [Wilks, 2006] providing an estimation of its added value. Added value is often more important than a measure of skill [Jolliffe and Stephenson, 2003].

The objective of this work is to assess the potential of NWP for flood early warning purposes, and the possible improvement that bias correction can provide. The study is focused on the comparison of streamflow forecasts obtained from the post-processed precipitation forecasts, particularly the comparison of ensemble forecasts and their potential in providing skilful food forecasts in a tropical mountainous area.

The study area is a páramo (tropical high mountain ecosystem) zone in Bogotá (Colombia), characterized by soils with a high water storage capacity and high conductivity with a hydrologic behaviour for which still major gaps in knowledge exist [Buytaert et al., 2006a, 2005b, Reyes, 2014, Sevink, 2007] and were the hydrometeorological data are scarce.

In Colombia, NWP models forecasts are provided by the Instituto de Hidrología, Meteorología y Estudios Ambientales (IDEAM), the national hydrometeorological institute. The raw output of these models is deterministic. In this study, the WRF model operated by IDEAM is used to produce precipitation forecasts that are post-processed and used to drive a hydrologic model of the páramo area. The discharges obtained from the hydrological model are used to assess the skill of the WRF model.

6.2 Methods and data

6.2.1 Study Area

The Tunjuelo river basin is located in the south of the city of Bogotá (see Figure 6.1). Its area is approximately 380 km^2 and the upper basin corresponds to a páramo area. The upper basin is composed of three watersheds, Chisaca, Mugroso and Curubital (see Figure 6.1-c). These discharge into two reservoirs (Chisaca and Regadera) with volumes of 3.3 Mm^3 and 6.7 Mm^3. The reservoirs are operated to supply 1.2 m^3/s of water to the south of Bogotá. Flood waves in the urbanized lower basin are dominated by the discharge release of the two reservoirs.

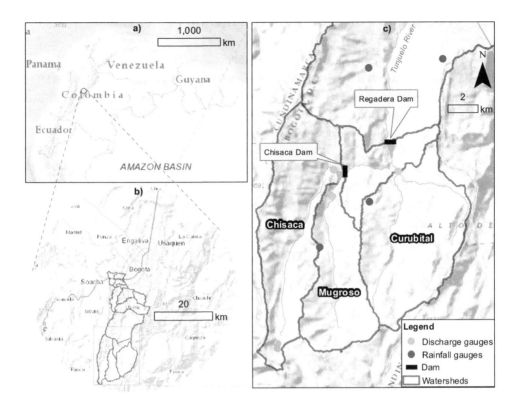

FIGURE 6.1: Study area. Service Layer Credits: Esri, DeLorme, NAVTEQ, TomTom, Intermap, increment P Corp., GEBCO, USGS, FAO, NPS, NRCAN, GeoBase, IGN, Kadaster NL, Ordnance Survey, Esri Japan, METI, Esri China (Hong Kong), swisstopo, and the GIS User Community

The last major flood event in the basin occurred in 2002, causing the river to change its course to flow into two mining pits that since act as inline reservoirs. The peak release of the Regadera dam for that event reached 100 m^3/s [Rogelis, 2006], which caused flooding downstream. The current flood warning criteria include warning levels set on the water levels in the Regadera Reservoir. Forecasting of the input discharges to this reservoir, as well as to the Chisaca reservoir, which is located immediately upstream, is crucial.

The upper basin of the Tunjuelo river has a unimodal precipitation regime (rainy season April-November) [Bernal et al., 2007]. The largest discharges in the upper Tunjuelo basin occur during the months May, June and July [Rogelis, 2006]. The monitoring network is shown in Figure 6.1. Two tipping bucket telemetric rain gauges currently operate in the upper Tunjuelo river basin and three discharge gauges are available in the three watersheds that discharge into the two reservoirs of the upper basin with hourly records.

6.2.2   WRF model data and observed rainfall fields

The hydrometeorological agency of Colombia, IDEAM runs the Weather Research and Forecasting (WRF) model version WRFV3.1 for Bogotá, using the initial conditions provided by the GFS model [Ruiz, 2010]. The WRF model has been used at IDEAM since 2007 to produce forecasts at national level, as well as in the Bogotá region at higher resolution [Arango and Ruiz, 2011].

The data set used for this study corresponds to 107 selected days when significant storms were recorded. For each of these days, forecasts with the WRF model were generated at 00:00 GMT, 06:00 GMT, 12:00 GMT and 18:00 GMT. This choice of storms ignores days when the WRF model would have delivered false-alarms, focusing exclusively on the days when high precipitation was effectively measured.

The storms typically occur in the afternoon, thus the WRF forecast have a lead time in a range between 0 and 18 hours. The simulation comprised the three nested domains, centred in Bogotá. The coarsest domain covers most of the Colombian territory with a spatial resolution of 15 km; the intermediate domain covers mainly the central and eastern cordilleras with a spatial resolution of 5 km; and the finest domain covering Bogotá only has a spatial resolution of 1.67 km [Arango and Ruiz, 2011]. The parameterisation of the model corresponds to that used by IDEAM for its routinely forecasts [Arango and Ruiz, 2011]. The lead time of the WRF model used operationally is 72 hours, though in this study a lead time of 48 hours was used

Observed hourly precipitation data was available from the tipping bucket rainfall gauges. These were used to produce rainfall fields through inverse distance weighing interpolation (IDW).

Both WRF forecast rainfall fields and IDW rainfall fields were transformed to time series of mean average precipitation for the Chisacá, Mugroso and Curubital watersheds (see Figure 6.1) that correspond to the páramo area of the Tunjuelo river basin.

6.2.3   Methodology

The hydrology of the Páramo watersheds of the Tunjuelo river was modelled using the TOPMODEL [Beven and Kirkby, 1979]. This was calibrated and compared with other two models in Chapter 5 and found to provide a realistic representation of hydrological processes and good performance in the páramo watersheds. Furthermore, the preference of TOPMODEL for páramo areas has been reported previously by other authors [Buytaert and Beven, 2011]. The hydrological models were driven with precipitation input obtained from inverse distance interpolation (IDW) of hourly rainfall obtained from gauges up to the time of start of the forecast (T0). From T0 the model was driven using rainfall forecasts corresponding to: a) Zero rainfall forecasts; b) raw forecasts from the WRF model; c) bias corrected WRF forecasts; d) and precipitation forecast ensembles obtained from the post processing of the WRF model.

Both precipitation forecasts and streamflow forecasts were verified through the use of skill scores and rank histograms. In the case of streamflow forecasts obtained from the WRF model, these were compared with a reference time series to assess its added value. The procedures for the production of the different rainfall forecasts and the verification process are described in the following sections.

6.2.3.1 Generation of Precipitation Forecasts

The precipitation forecast to drive the models were generated under four strategies: a) Zero rainfall forecasts: After T0 values of zero precipitation are used to drive the TOPMODEL. b) Raw forecasts from the WRF: the only pre-processing of the WRF forecasts is the sampling of the grids to obtain the hourly mean areal precipitation for each watershed. No post-processing is applied. c) Bias correction of WRF: The time series of mean areal precipitation obtained from the WRF model are bias corrected through Distribution-Based Scaling - DBS [Yang et al., 2010]. As reference values the time series of mean areal precipitation obtained from (IDW) are used and the correction is carried out for each lead-time of the WRF model. The Distribution-Based Scaling (DBS) approach uses two steps: (1) Correction of the percentage of wet time steps. A precipitation threshold of 0.1 mm (rainfall gauge accuracy) was used, below which a time step is considered to be dry and (2) Transformation of the remaining precipitation to match the observed frequency distribution. The gamma distribution is used to describe the probability distribution function (PDF) of precipitation intensities given its ability to represent the asymmetrical and positively skewed distribution of precipitation. The distribution parameters are estimated using maximum likelihood estimation (MLE). To capture the main properties of normal precipitation as well as extremes, the precipitation distribution is divided into two partitions separated by the 95th percentile. The resulting distribution corresponds to a double gamma distribution. The two sets of parameters are used to correct the WRF model outputs. d) Bias correction and generation of ensembles through post processing of rainfall forecast from the WRF model: The reference and forecast time series were organized according to lead time and a two stage post processing model was applied to reflect the intermittent nature of rainfall [Clark et al., 2004, Rene et al., 2012]. The first stage corresponds to the probit model [Bremnes, 2004, Kleiber et al., 2012, Scardovi, 2015] to simulate the occurrence of precipitation, and the other corresponds to the amount of precipitation, given occurrence, for which quantile regression was used. The mean precipitation time series are first disaggregated into a time series of occurrence (1 = wet time step and 0= dry time step) and precipitation amounts (only wet time steps). The time series of occurrence is used as the response variable for the probit regression model, and the time series of precipitation amounts is used as the response variable for the quantile regression model.

In the probit regression model, given binary observations of precipitation occurrence $y_1,...y_n$, and $x_{i1},...x_{ik}$ covariates associated with the ith response, the probability that $y_i = 1$, p_i, is written as [Albert, 2009]:

$$p_i = P(y_i = 1) = \Phi(x_{i1}\beta_1 + ... + x_{ik}\beta_k) \tag{6.1}$$

where $\beta = (\beta_1,...,\beta_k)$ is a vector of unknown regression coefficients and $\Phi()$ is the cdf of a standard normal distribution. If we place a uniform prior on β, then the posterior density is given by:

$$g(\beta \mid y) \propto \prod_{i=1}^{n} p_i^{y_i}(1-p_i)^{1-y_i} \tag{6.2}$$

The binary response corresponds to the occurrence $y_i = 1$ or non-occurrence of rain $y_i = 0$. A latent variable Z_i is introduced in such a way that if Z_i is positive for $y_i = 1$ and Z_i is negative for $y_i = 0$. This latent variable is related to the k covariates by the normal regression model:

$$Z_i = x_{i1}\beta_1 + ... + x_{ik}\beta_k + \epsilon_i \tag{6.3}$$

where ϵ_1 , ..., ϵ_n are a random sample from a standard normal distribution. Then:

$$P(y_i = 1) = P(Z_i > 0) = \Phi(x_{i1}\beta_1 + ... + x_{ik}\beta_k) \tag{6.4}$$

This can be considered as a missing data problem where there is a normal regression model of latent data $Z_1,...Z_n$ and the observed responses are missing or incomplete and it can only be observed if $Z_i > 0(y_i = 1)$ or if $Z_i \leq 0(y_i = 0)$.

In order to avoid distributional assumptions, quantile regression (QR) will be used to describe the probability of the amounts of precipitation given its occurrence [Bremnes, 2004]. Let $r_1,...r_{n*}$ denote observed precipitation amounts of cases with observed precipitation above a given lower threshold and $z_1,...z_{n*}$ corresponding predictor values where $z = (z, ..., z)^T$. For linear quantile functions [Bremnes, 2004]:

$$q_\theta(z_i; \beta) = \beta_0 + \sum_{k=1}^{K} \beta_k z_{ik} \tag{6.5}$$

an estimate of the $q_\theta(z_i; \beta), 0 < \beta < 1$ quantile, is obtained by solving the following minimization problem with respect to β :

$$\arg\min_{\beta} \sum_{i=1}^{n*} p_\theta \left(r_i - q_\theta \left(z_i; \beta \right) \right) \tag{6.6}$$

where the function $q_\beta(*)$, is defined in terms of the absolute deviation of residuals (u) by:

$$p_\theta(u) = \begin{cases} u\theta & if \ u \geq 0 \\ u(\theta - 1) & otherwise \end{cases} \tag{6.7}$$

If several quantiles are of interest, the minimization must be repeated for each quantile. A potential problem with using QR for the derivation of multiple conditional quantiles is that quantiles may cross, yielding predictive distributions that are not monotonously increasing, as a function of increasing quantiles [López López et al., 2014]. In the present research study, the technique proposed by Muggeo et al. [2013], and implemented in the package quantregGrowth developed in the R environment [R Development Core Team, 2010] is used. This technique estimates nonparametric growth charts via quantile regression. Quantile curves are estimated via B-splines with a quadratic penalty on the spline coefficient differences, and non-crossing and monotonicity restrictions are set to obtain plausible estimates.

Two configurations were tested: a) Quantile regression applied to the raw precipitation data; b) Quantile regression on the data transformed into normal domain through normal quantile transformation (NQT) [Bogner et al., 2012]. For configuration b) the time series of observed precipitation and forecast precipitation are transformed into the normal domain. After the derivation of the quantiles, the variables are back-transformed into original space. The rationale for using the transformation is that the joint distribution of transformed time series appears to be more linear, and can thus be better described by linear conditional quantiles [López López et al., 2014].

Back-transformation is, however, problematic if the quantiles of interest lie outside of the range of the empirical distribution of the untransformed variable in original space. To address this issue linear extrapolation in the tails of the distribution was used [López López et al., 2014].

To generate the probabilistic forecast of precipitation occurrence, the latent variable in the probit model is first sampled. The random value u is sampled from a uniform distribution, If $Z_i < u$ no precipitation occurs, if $Z_i \geq u$ precipitation is set to occur and the amount is computed with the quantile regression model. Given the occurrence of precipitation, the quantile regression model corresponding to the lead-time under consideration is used to obtain the quantiles of the conditional distribution. The inverse cumulative function is approximated by a cubic spline and subsequently 50 uniform random numbers in the range 0-1 are generated to sample corresponding precipitation values from the inverse cumulative function.

6.2.3.2 Verification of forecasts

Deterministic forecasts of precipitation and discharge were verified using the Mean Absolute Error (MAE), the Mean Squared Error (MSE), the Mean Error (ME) and the skill score (SS) based on the MSE using as reference forecasts provided by climatological values (subscript Clim in Equation 6.11 and Equation 6.12) [Wilks, 2006] and zero forecasts (subscript 0 in Equation 6.13 and Equation 6.14). The corresponding formulas are shown in Equation 6.8 to Equation 6.14, where (y_k, o_k) is the k_{th} of n pairs of forecast and observed values at a particular lead time. In the case of discharge, the discharge forecast obtained from the hydrological model driven with a precipitation input equal to zero, was used for comparison with the discharges obtained from the WRF forecasts.

$$MAE = \frac{1}{n} \sum_{k=1}^{n} |y_k - o_k| \tag{6.8}$$

$$MSE = \frac{1}{n} \sum_{k=1}^{n} (y_k - o_k)^2 \tag{6.9}$$

$$ME = \frac{1}{n} \sum_{k=1}^{n} (y_k - o_k) = \bar{y} - \bar{o} \tag{6.10}$$

$$MSE_{clim} = \frac{1}{n} \sum_{k=1}^{n} (\bar{o} - o_k)^2 \tag{6.11}$$

$$SS_{Clim} = \frac{MSE - MSE_{Clim}}{0 - MSE_{Clim}} = 1 - \frac{MSE}{MSE_{Clim}} \tag{6.12}$$

$$MSE_0 = \frac{1}{n} \sum_{k=1}^{n} (0 - o_k)^2 \tag{6.13}$$

$$SS_0 = \frac{MSE - MSE_0}{0 - MSE_0} = 1 - \frac{MSE}{MSE_0} \tag{6.14}$$

In order to assess the performance of the probit model constructed to model the occurrence of precipitation from the WRF model, the ROC (Relative Operating Characteristic) diagram was used. This is a discrimination-based graphical forecast verification display [Wilks, 2006]. For perfect forecasts the ROC curve consists of two line segments coincident with the left boundary and the upper boundary of the ROC diagram. At the other extreme of forecast performance, random forecasts consistent with the sample climatological probabilities, the ROC curve will consist of the 45 degrees diagonal connecting the points (0, 0) and (1,1). ROC curves for real forecasts generally fall between these two extremes, lying above and to the left of the 45 degrees diagonal. Forecasts with better discrimination exhibit ROC curves approaching the upper-left corner of the ROC diagram more closely, whereas forecasts with very little ability to discriminate the event exhibit ROC curves very close to the diagonal [Wilks, 2006]. To summarize the ROC diagram the area under the ROC curve was used [Wilks, 2006].

Regarding the quantile regression models, the Pseudo R-Square measure [Koenker and Machado, 1999] was used to compare the models with normal quantile transformation and the ones based on raw data. The Pseudo R-Square measure was proposed by Koenker and Machado [1999] as a goodness of fit indicator for quantile regression by comparing the sum of squared distances for the model of interest with the sum of squared distances between the observed and the fitted values that would be obtained if only the intercept term is included in the model [Hao and Naiman, 2007].

Ensemble forecasts of both precipitation and discharge were verified using Rank Histograms. Rank histograms allow evaluating whether a collection of ensemble forecasts satisfy the consistency condition [Wilks, 2006]. Rank histograms are constructed by accumulating the number of cases over space and time when the verifying analysis falls in any of m+1 intervals, where each of the m +1 intervals is defined by an ordered series of m ensemble members, including the two open ended intervals [Jolliffe and Stephenson, 2003]. Reliable or statistically consistent ensemble forecasts lead to a rank histogram that is close to flat [Jolliffe and Stephenson, 2003].

In order to assess the probabilistic forecasts of discharge obtained from the WRF precipitation ensembles, the Continuous Ranked Probability Skill Score (CRPSS) was used. The CRPSS is given by [Alfieri et al., 2014, Hersbach, 2000]:

$$CRPSS = \frac{\overline{CRPS_{ref}} - \overline{CRPS_{forecast}}}{\overline{CRPS_{ref}}} \tag{6.15}$$

where:

$$CRPS = \int_{\infty}^{-\infty} [F(y) - F_0(y)]^2 \, dy \tag{6.16}$$

and

$$F_0(y) = \begin{cases} 0, & y < observed \ value \\ 1, & y \geq observed \ value \end{cases} \tag{6.17}$$

$F(y)$ is a stepwise cumulative distribution function (cdf) of the ensemble of each considered forecast. CRPSS ranges between 1 and $-\infty$. Forecast ensembles are only valuable when $CRPSS > 0$, while a

CRPSS of 1 indicates a perfect forecast. This is when the forecasts perform better than the reference. The discharge obtained from forecast precipitation equal to zero will be used as reference. Since this is a deterministic forecast, the $CRPS_{ref}$ corresponds to the mean absolute error of all forecast and observed pairs (Hersbach 2000).

6.3 Results

6.3.1 Bias correction of precipitation forecasts through DBS

Figure 6.2 shows the empirical cumulative distribution functions (ECDF) of the raw WRF data, the observed precipitation obtained from IDW interpolation, and the bias corrected WRF data for the Mugroso watershed, results for the other two watersheds were similar and not shown here for brevity. The raw WRF and IDW time series were obtained from the corresponding rainfall fields as the mean areal precipitation calculated in the three watersheds of analysis by sampling the raster layers with the polygons of the watersheds.

The behaviour of the ECDF for the other watersheds is similar. The ECDF of raw WRF precipitation shows that both overestimation and underestimation of precipitations occurs. Underestimation is very noticeable for lead times up to 2 hours for values in the upper percentiles. In general for lead times of 3 hours and higher, overestimation of high precipitation values occurs.

6.3.2 Quantile regression model

Figure 6.3 shows the Pseudo R-Square for all quantile regression models for the Mugroso watershed and for the coarsest domain (resolution of 15 km), the results are similar for the other watersheds and domains. The R-Square values are pooled for lead times up to 12 hours. 12 hours was chosen as a first estimate of the usable lead-time for the forecast, since NPW models describing local-regional convective systems and orographic processes have been observed to significantly decrease their skill at lead times beyond 12-48 hours [Cuo et al., 2011]. The figure shows that the values of the Pseudo R-Square up to the 50th percentile are higher when the normal quantile transformation is used, thus indicating a better fit of the quantile regression model. For higher percentiles the Pseudo R-Square of the models with normal quantile transformation is in approximately in the same range of the raw data or lower. This means that for percentiles above the 50th percentile the fit of the quantile regressions is better for the raw data than for the normal transformed data.

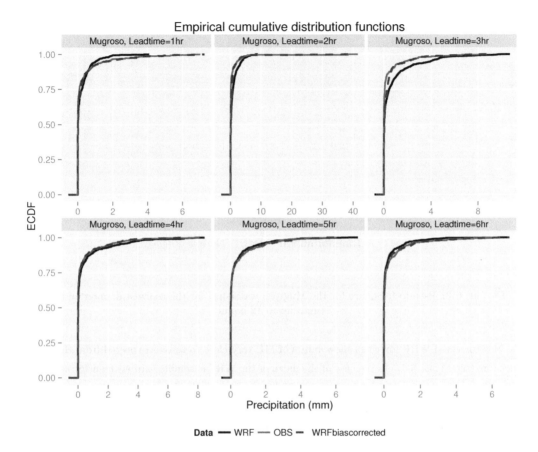

FIGURE 6.2: Empirical cumulative distribution function (ECDF) for the Mugroso watershed hourly precipitation for lead times up to 6 hour

6.3.3 Verification of precipitation forecasts

6.3.4 Verification of deterministic precipitation forecasts and ensemble mean

Figure 6.4 shows the performance measures showed in Equation 6.8 to Equation 6.14. The ME of the raw WRF forecasts shows that in the three catchments (Chisaca, Mugroso and Curubital) for lead times between 1 and 5 hours, there is over prediction (positive ME), while after 5 hours the ME shifts to negative values indicating underprediction.

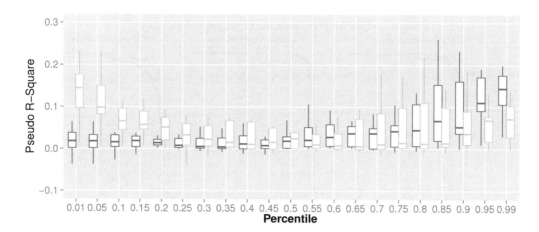

FIGURE 6.3: Pseudo R-Square for the Mugroso watershed for the coarsest domain and lead times up to 12 hours

The bias corrected WRF forecasts show values of ME very close to zero as expected from the applied procedure (see Figure 6.4). In the case of the mean of the WRF ensemble without transformation, the ME is close to zero manly for lead times up to 6 hours. Beyond this lead-time there is a tendency to underpredict precipitation. The behaviour of ME for the mean of the ensemble with normal quantile transformation is very similar to when the transformation is not applied.

The highest values of MAE (see Figure 6.4) are obtained from the DBS bias corrected forecast for lead times larger than 6 hours, for shorter lead times the raw WRF forecasts exhibit the highest values. In the case of the MSE (see Figure 6.4) the highest values are obtained from the bias-corrected WRF forecasts at lead times of 2 and 8 hours, when all forecasts show peak values. The increment in MSE when the bias correction is applied is mainly due to the influence of high precipitation values that are not forecast by the WRF model. These intense precipitation missing forecasts (mainly present at lead times of 2 and 8 hours) cause that the WRF values higher than the 95th percentile significantly increase with the bias correction procedure, increasing their square difference with the observed value.

For the MSE and MAE, the best performance is obtained from the ensemble mean of the WRF forecasts, with and without normal quantile transformation (see Figure 6.4). These provide very similar performance. Regarding the skill score based on the MSE using the mean value of the time series as reference (see Equation 6.11 and Equation 6.11) negative values are obtained for all forecasts, with the bias corrected WRF and the raw WRF forecasts showing the worst skill. However, in the case of the ensemble mean, both quantile transformed and raw, the values are very close to zero albeit

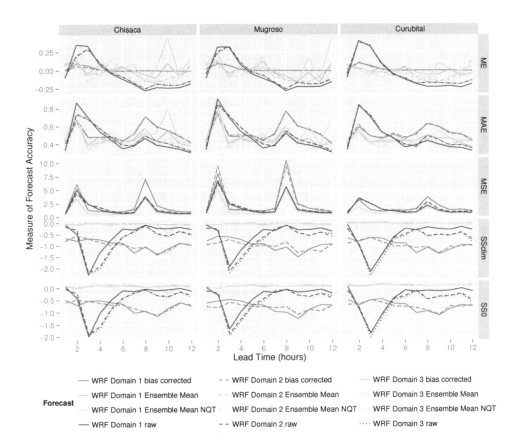

FIGURE 6.4: Accuracy Measures for deterministic precipitation and ensemble mean obtained from the WRF model . Domain 1 , domain 2 and domain 3 correspond respectively to the domains with resolutions 15 km, 5 km and 1.67 km

negative. When zero precipitation is used as reference, the behaviour of the score is very similar to SSclim. However, most values for the ensemble mean forecast are positive both with raw and with normal quantile transformed data, with a maximum value of 0.16. Thus, the ensemble mean forecasts provide the best results in comparison to the other forecasts. In the case of the mean of the time series as reference, they are approximately as good as the reference and compared to zero precipitation as reference; they provide an improvement up to 16% over the reference.

Given that one storm present at lead times of 2 and 8 hours, has a significant influence in the results shown in Figure 6.4, the accuracy metrics were recalculated from the data set excluding this storm, which represents the highest precipitation value that was missed by the WRF model. The results are shown in Figure 6.5. In this, there is a reduction of the ME and MAE that is highly noticeable in all

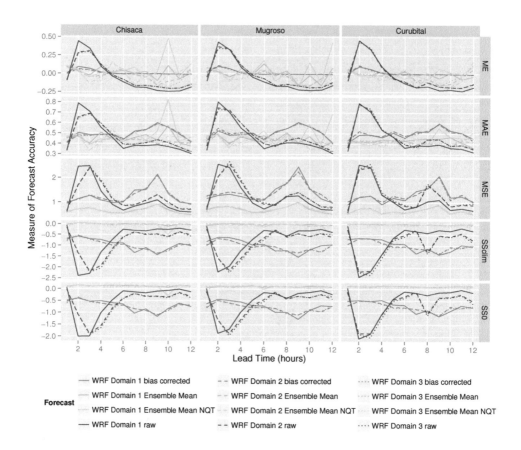

FIGURE 6.5: Accuracy Measures for deterministic precipitation and ensemble mean obtained from the WRF model without the highest precipitation missed by the WRF model. Domain 1, domain 2 and domain 3 correspond respectively to the domains with resolutions 15 km, 5 km and 1.67 km

the forecasts. The MSE reduces as well (see Figure 6.4 in comparison to Figure 6.5), and the lowest values are exhibited by the ensemble mean of both raw and normal quantile transformed data, which show a very similar behaviour. SSclim and SS0 (see Figure 6.5) behave very similarly to the case were all the storms are included showing approximately the same values as the analysis with all the storms.

6.3.5   Verification of deterministic discharge forecasts and ensemble mean

The results for discharge using the precipitation forecasts are shown in Figure 6.6. The ME shows that the raw WRF forecasts produce an overestimation of discharges up to a lead time of 8 hours, while for larger lead times underestimation occurs. When the bias corrected WRF forecasts are used, the ME is close to zero, slowly increasing with lead-time. The use of the mean of the ensembles with and without normal quantile transformation produces ME close to zero up to lead times of 4 hours, beyond which underestimation occurs (negative values of ME). In terms of the MSE the mean of the ensemble with and without normal quantile transformation produce the smallest values for all lead-times, which is reflected in the SSclim, where they produce the highest values, albeit very close to the values obtained from using a zero precipitation series as forecast.

Figure 6.7 shows the performance measures calculated from the stream flow simulations excluding the storm with the highest precipitation missed by the WRF model. The performance measures show a significant improvement in comparison with Figure 6.6, particularly for the DBS bias corrected data. However, the values of SSclim, continue to be very close to the values obtained from using zero precipitation as forecast.

6.3.6   Verification of probabilistic forecasts

Figure 6.8 shows the area under the ROC curve (AROC) for the páramo watersheds (Chisaca, Mugroso and Curubital in Figure 6.2) , for the three domains of the WRF model for lead times up to 12 hours and for a precipitation forecast of zero. The behaviour of the AROC is similar in the three watersheds, the highest values are obtained in the first two hours, dropping afterwards and reaching values close to 0.5 after lead times of 5 hours. The AROC values for a forecast of zero precipitation is very close to 0.5 for all lead times, thus is almost equal to random forecasts consistent with the sample climatological probabilities [Wilks, 2006]. This shows that the forecast has some skill at short lead times in comparison to both zero precipitation forecasts and random forecasts, although that skill is limited.

The rank histograms of the precipitation ensembles are shown in Figure 6.9. The behaviour for all domains and all lead times is similar (Figure 6.9 shows lead-times up to six hours corresponding to the left column of the graph). The rank histogram is approximately uniform from rank 1 to rank 5 and from rank 6 to rank 10 there is an increase of frequency of the IDW precipitation values falling in these higher ranks. A similar behaviour in all rank histograms was observed when the storm with the highest precipitation missed by the WRF model was excluded. This suggests that the WRF ensemble is slightly underforecasting.

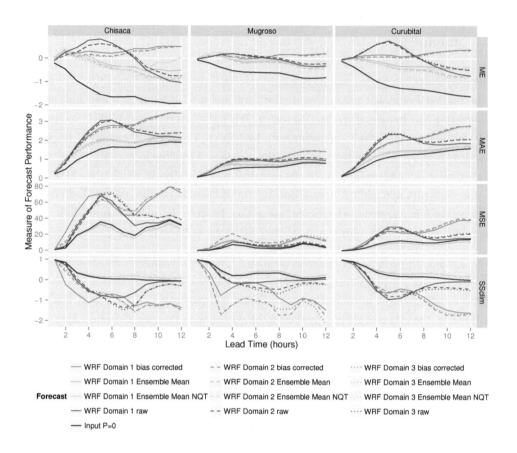

FIGURE 6.6: Performance Measures for deterministic discharge and ensemble mean forecasts. Domain 1 , domain 2 and domain 3 correspond respectively to the domains with resolutions 15 km, 5 km and 1.67 km

Figure 6.10 shows the rank histograms for the discharges obtained from the precipitation ensemble using the TOPMODEL. The rank histograms show approximate uniformity mainly for lead times up to 3 hours. For longer lead times a slight under dispersion is observed.

Figure 6.11 shows the CRPS for the discharge ensembles, the MAE for the forecast produced by using zero as precipitation forecast and the corresponding CRPSS. The CRPSS exhibits values in the range 10.5-21% (see Figure 6.11) with most values in a range of 14-16%. The same behaviour is observed in the three watersheds. The comparison between normal quantile transformed quantile regressions and the quantile regressions with raw data again shows that the difference between the two is negligible.

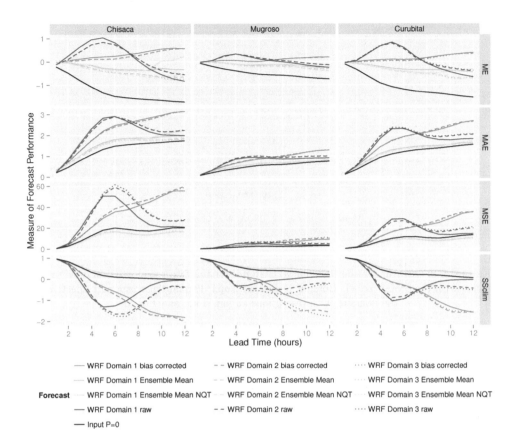

FIGURE 6.7: Performance Measures for deterministic discharge and ensemble mean forecasts obtained excluding the storm with the highest precipitation missed by the WRF model. Domain 1 , domain 2 and domain 3 correspond respectively to the domains with resolutions 15 km, 5 km and 1.67 km

6.3.7 Discussion

6.3.7.1 Evaluating precipitation forecasts from the WRF model

The comparison of ECDF of the raw WRF forecasts and the ECDF of the precipitation obtained from IDW interpolation shows that the WRF model tends to over predict precipitation. This behaviour has been observed in other tropical areas [Mourre et al., 2015]. The details of the implementation of the WRF model are out of the scope of this study. However, possible causes of the precipitation error found by other authors include: errors in the lateral boundary conditions [Ochoa et al., 2014]; poor

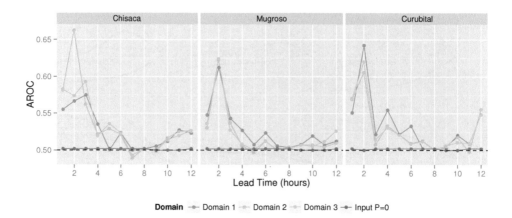

FIGURE 6.8: Area under the ROC curve (AROC) for the forecast of occurrence of precipitation with the probit model. Domain 1, domain 2 and domain 3 correspond respectively to the domains with resolutions 15 km, 5 km and 1.67 km

representation of the topography [Ochoa et al., 2014]; and choice of convective treatment, microphysics and planetary boundary layer [Jankov et al., 2005].

Figure 6.12 shows the precipitation obtained from IDW interpolation compared with the precipitation forecast of the WRF model for the highest resolution domain (1.67 km) and all the storms considered in this analysis. It can be observed that there is a lack of correlation between both data series. Most WRF values are higher than the IDW values, and there are individual cases where the WRF model did not detect the occurrence of high precipitation. Missing or underpredicted forecasts of intensive precipitation have also been observed by other authors [Kryza et al., 2013, Liu et al., 2015], which may suggest the need for an improvement in model parametrization and data assimilation.

No significant differences among domains are observed in the behaviour of the scores for the tests carried out in this study. Other applications of NWP (e.g. Roberts et al. [2009]) show that finer resolutions are capable of producing more accurate predictions, and that physics configuration, resolution and initial conditions highly influence the WRF model performance [Kryza et al., 2013]. The similarity of results regardless of resolution found in this study may be also related to parametrization deficiencies, or to inability to sufficiently resolve the topography in the models. A more detailed review of the WRF model would be required to reveal possible deficiencies, and suggest improvements.

The WRF model has shown to be highly sensitive to the parametrization of cumulus and microphysical processes [Rama Rao et al., 2012, Remesan et al., 2014]. Parameter sensitivity is generally dependent on local conditions [Di et al., 2014], and the best configuration varies with time and rainfall threshold

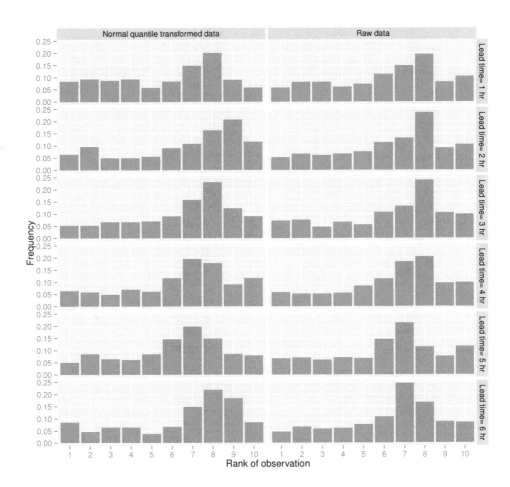

FIGURE 6.9: Rank histograms for the WRF ensemble for the Mugroso watershed for the finest domain (resolution of 1.67 km) and lead times up to 6 hours

[Jankov et al., 2005]. Therefore, a step forward would be to test various model physical parametrizations assessing their sensitivity and search for an optimum or a set of optimum configurations that help to improve the prediction of convective intensive precipitation. The use of observing systems such as radars, gauges and satellites for data assimilation in NWP analysis [Cuo et al., 2011, Liu et al., 2015, Rossa et al., 2011, Yucel et al., 2015].

The application of the DBS approach has been reported previously to considerably reduce the differences in rainfall frequency between the observations and forecasts [Yang et al., 2010]. This reduction of differences was found in this study, as shown in Figure 6.12. However, in the time series for lead times where high precipitation values were observed but not forecasted, the bias correction had the effect of reducing performance, particularly when measured by the MSE, since this is more sensitive to

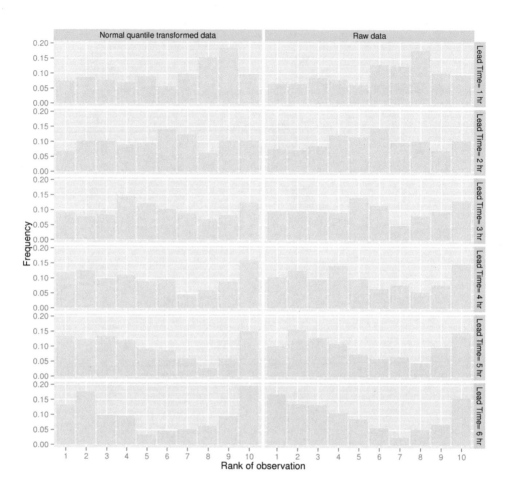

FIGURE 6.10: Rank histograms for the discharge ensemble using as reference time series the discharge simulated with the TOPMODEL

large residuals. This behaviour suggests that in these time series the DBS bias correction introduced undesired effects, which limit its effectiveness [Ehret et al., 2012].

A similar behaviour was observed with the mean of the ensemble with and without normal quantile transformation, albeit with less impact on the performance measures in comparison to the DBS approach. High observed non-forecast values of precipitation caused a reduction in performace reflected in higher ME, MAE and MSE in comparison to the same metrics calculated with a data set excluding the highest storm that was missed by the WRF model. No significant difference exists between the ensemble means obtained form raw and normal quantile transformed data.

These biases are the result of the inability of the WRF model to adequately represent convective

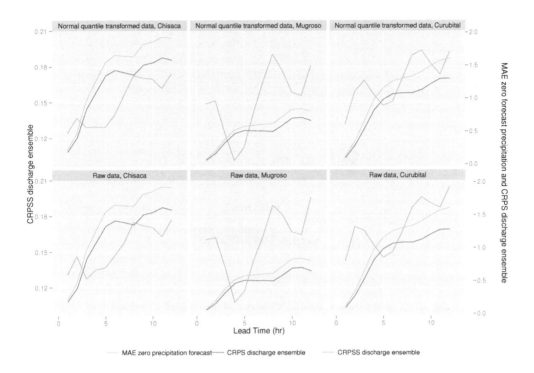

FIGURE 6.11: CRPSS for the discharge ensembles

precipitation, this is, the failure to correctly predict when large events occur. This is a behaviour observed by other authors. For instance, [Verkade et al., 2013] found that post-processing does not improve on all qualities at all lead times and at all levels of the verifying observations. The cause is rooted in the impossibility of the post-processing approaches to replace adequate model representation of physical processes [Haerter et al., 2011]. As stated by [Haerter et al., 2011] the conditions on climate model data to make the application of statistical bias correction schemes reasonable are that it must be ensured that the model provides a realistic representation of the physical processes involved; and that the quantitative discrepancies between the modelled and observed probability density function of the quantity at hand must be constant in time. Similarly, Chen et al. [2013] emphasize the impossibility of success of any bias correction method if there is no coherence between simulated and observed precipitation. The WRF data used in this study shows limitations in fulfilling both these conditions, which leads to the result that the post-processed precipitation is no more skilful than the sample climatology and provides only a modest improvement in comparison to the zero precipitation forecast. However, an important result of this analysis is that despite the limitations, the ensemble mean outperforms the DBS bias correction and seems to be less sensitive to the presence of intensive

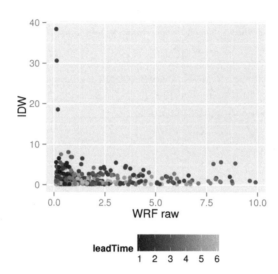

FIGURE 6.12: IDW precipitation for the Chisaca, Mugroso and Curubital watersheds vs WRF raw precipitation for the highest resolution domain (1.67 km)

precipitation that is not forecast by the WRF model. The post-processing to generate the ensembles is more sophisticated than just correcting the mean bias [Robertson et al., 2013]. During the process, different distributions are used for each raw WRF forecast which results in an improvement of skill that outperforms the bias correction through DBS. The magnitude of the bias of the ensemble mean is nearly always smaller than the raw forecasts and the DBS bias corrected forecasts, particularly for lead times up to 6 hours. And shows a superior behaviour in terms of skill scores.

Regarding the verification of the precipitation ensembles, the rank histograms show that the IDW precipitation falls too often between the ranks 6 and 10. This reflects overdispersion of the ensemble towards low values. This produces an underforecasting bias in the ensemble. The same behaviour of the rank histograms is observed in the ensembles obtained from quantile regression when no transformation is used in the precipitation data and when raw data is used. This means that the transformation of the values does not improve the consistency of the ensemble.

The AROC of the occurrence of precipitation shows that values higher than 0.5 are obtained in all watersheds for lead times up to 5 hours. For longer lead times the probit model does not have the ability to discriminate between the occurrence or non-occurrence of precipitation. In all domains and watersheds, the highest values of AROC are obtained for lead-times between 2 and 3 hours. The highest values of AROC are in the range of 0.67-0.63. This means that there is some capacity of the probit model to forecast occurrence of precipitation for lead times up to 5 hours. For longer times the model does not have any skill.

The lead time for which a reduction in skill is observed is consistent with other studies where the performance significantly reduces after a few hours [Liu et al., 2015], implying that the current ability of NWP in tropical mountainous areas still provides a limited extension of lead time beyond the concentration time of a watershed.

6.3.7.2   Evaluating discharge forecast

The comparison of Figure 6.6 and Figure 6.7 shows that the biases in the precipitation forecasts reflect in the biases of the discharge. When the single event with very high precipitation that is not forecast by the WRF is excluded from the analysis, the reduction of bias in the precipitation is observed as well as the reduction in the bias of the discharge. This is consistent with the findings of other studies where the errors in bias corrected precipitation lead to amplified errors in modelled runoff [Teng et al., 2015], since in this analysis the increase in errors in the bias corrected time series amplify the errors in the simulated discharges.

The forecast generated with the ensembles exhibit higher skill than the deterministic forecasts (based on raw, bias corrected and zero precipitation). The higher skill of the ensemble mean in comparison with deterministic forecasts has been also found in other flood forecasting systems driven by the WRF model [Calvetti et al., 2014] and other NWP models [Vincendon et al., 2011].

In terms of lead-time, a fist limit is provided by the skill of the WRF model to forecast the occurrence of precipitation that is 5 hours (see Figure 6.8). A limit is also observed in the skill scores of the discharge forecast in Figure 6.7 and Figure 6.7. After a lead-time of 2 hours (approximate time of concentration of the watersheds), the skill (SSclim in Figure 6.6 and Figure 6.7) drops significantly. For the discharge obtained from raw and DBS bias corrected precipitation, three and four hours are the corresponding lead times for which added value is totally lost in comparison with the climatology (SSclim in Figure 6.6 and Figure 6.7 reaches zero). In the case of the ensemble means after two hours the skill decreases progressively up to a lead-time of 6 hours. Beyond this lead-time, the values remain approximately constant in a range of 0.2 to 0.5 for the three watersheds. This behaviour indicates that the use of the ensemble mean provides a lead time of 6 hours, which is three hours longer in comparison to the lead time provided by the raw WRF forecast or the DBS bias corrected WRF forecast. A lead time of 6 hours is consistent with other flood forecasting systems in small mountainous catchments driven by NWP models [Verkade and Werner, 2011].

The hydrological model produces ensemble results with rank histograms that do not reflect the underforecasting of the precipitation ensemble, with approximately uniform rank histograms mainly for the first lead times. Approximately the same shape of the rank histograms is observed for ensembles obtained from quantile regression when no transformation is used in the precipitation data and when raw data is used. The minor influence of the transformation is consistent in all the performance assessments. This may be due to the relatively low improvement of the fit of the quantile regressions when the normal quantile transformation is used. According to Figure 6.3, the most significant improvement is found in the low percentiles (up to the 25th percentile). For higher percentiles, the improvement is not very significant and for some percentiles degradation of goodness of fit occurs. As floods occur for higher rainfall percentiles, the similarity in behaviour is logical.

The results of the CRPSS show that that driving the hydrological model with WRF ensembles improves the forecasts in a range of 8.5-22% in comparison with a forecast produced driving the hydrological model with zero precipitation for lead time between 1 and 12 hours. The positive values of the CRPSS imply added value to the forecasts, albeit modest. The CRPSS obtained from the ensembles is comparable to the CRPSS values found in other areas with other NWP models, albeit in the low range. E.g. Robertson et al. [2013] found a CRPSS of 37% on average for post processed ensembles in Australia, where rainfall is predominantly produced by large-scale synoptic systems that are better predicted by NWP models. Therefore, given the high complexity of the meteorological conditions of the study area, and despite the relatively poor skill of the WRF model in predicting precipitation amounts, the WRF model has shown promise at producing a benefit in its use for flood forecasting compared to not using precipitation forecasts. This is likely due to the skill found in the probit model in predicting the occurrence of rainfall.

Besides improvements in parametrization and data assimilation, as described previously, bias correction of stream flow could provide a further skill improvement [Yuan and Wood, 2012].

6.4 Conclusions

This chapter presents the assessment of WRF forecasts produced in a páramo area in Bogotá (Colombia). The WRF forecasts were used to drive a hydrological model. The simulated discharges were used to assess the value of the WRF forecasts for flood early warning. Results show that the streamflow forecasts obtained from a hydrological model driven by post-processed WRF precipitation add value to the flood early warning system when compared to zero precipitation forecasts. The WRF model for the study area provides forecasts that overpredict precipitation and that tend to fail to forecast high intensive precipitation. This behaviour may be due to parametrization deficiencies, errors in boundary conditions and poor representation of topography. There is a need for more detailed evaluation of the WRF model in this study area. The use of satellite and soon to be available radar data may improve performance. Furthermore, other convective and microphysics schemes should be assessed to identify the most suitable parametrization.

Bias correction through Distribution-Based Scaling - DBS significantly reduced the differences in rainfall frequency between observations and forecasts. However, a reduction of performance was observed when intensive precipitation, that is not forecast by the WRF is included in the data. Similar behaviour was observed with the mean of the ensembles with and without normal quantile transformation, albeit with less impact on the performance in comparison with DBS. This undesired behaviour is the result of the inability of the WRF model to adequately represent convective precipitation, which cannot be corrected through simple post-processing.

Despite the limitations in the WRF forecasts, the ensemble mean outperforms the DBS bias correction and seems to be less sensitive to the presence of highly intense precipitation that is not forecast by the WRF model. No precipitation forecast used in this analysis showed added value when compared to climatology. However, the reduction of biases obtained from the ensembles show potential of this method and model to provide usable precipitation forecasts. The probit model, used to forecast the precipitation occurrence based on the WRF forecast, showed that the WRF model has some skill at short lead times (up to 5 hours) in comparison to both zero precipitation forecasts and random forecasts, although that skill is limited.

Increases in precipitation biases are reflected in the discharge forecasts. However, discharge forecasts generated with ensembles exhibit higher skill than deterministic forecasts (based on raw, bias corrected and zero precipitation). Furthermore, the quality of these forecasts is better than what could be obtained using zero precipitation as input to the hydrological model.

The potentially usable lead times for forecasts obtained from the WRF model found in this analysis (5-6 hours) are in the range of lead times found in other studies with other NWP models. Despite the fact that the added value of the WRF model forecasts is modest, this shows promise for increasing forecast skill in areas of high meteorological and topographic complexity and the possibility of improvement.

Chapter 7

Conclusions and Recommendations

7.1   Conclusions

This dissertation addresses the risk knowledge and forecasting components of flood forecasting systems in developing cities, characterized by mountainous tropical environments. The research is framed in the holistic concept of people centred early warning, where integration of risk knowledge, monitoring and warning system, dissemination and communication and response capability determines the efficiency of early warning. The contributions of this research are intended to advance the knowledge required for design and operation of flood early warning in data-scarce watersheds from a hydrologic perspective, without neglecting the crosscutting nature of flood early warning in the flood risk management process.

The central question posed in the research was: *How can a reliable operational flood forecasting system be established in developing cities, considering uncertainty as an effective tool for decision making?*

In addressing the central question, a number of secondary questions were posed in order to focus on particular issues present in the issuing of warnings in developed cities: (i) prioritisation of watersheds to focalize flood early warning efforts; (ii) appropriate description of the spatial distribution of rainfall in areas with complex topography and meteorology; (iii) assessment of hydrological models in tropical high montane basins; and (iv) the potential use of numerical weather models for flood forecasting in tropical high montane basins.

Through the research of methods focused on the above subjects, this research hopes to contribute to closing the gap between developing and developed countries in methodologies applied to process, model and handle uncertainty in forecasts. The overall results show that innovative methods based on easily obtained data can be used efficiently to support regional risk management decisions, which constitute an important basis for flood early warning development. Furthermore, the development of flood early warning poses challenges in high montane basins that can be addressed through the application of methodologies that increase reliability in the models and through directing efforts to exploiting the potential that numerical weather models offer. However, there are limitations imposed not only by the lack of data characteristic of these areas, but also by the state of knowledge of the hydrologic processes of high montane basins and by the current capabilities of hydrologic and meteorological models.

The study area corresponds to the city of Bogotá (Colombia). Bogotá is located on a high plateau surrounded by hills in the Eastern Andes mountain range of Colombia. The western city limit is the Bogotá River, which drains a large plain called the Savannah of Bogotá. The regional flood risk analysis covered the mountainous area of the city and the precipitation analysis included a larger area including both the mountains and the urban plateau. Hydrological modelling was carried out in the Tunjuelo river basin, the largest tributary of the Bogotá River, located in the south of the city. This watershed is composed of a Páramo upper basin, a rural middle basin and an urban lower basin. The páramo upper basin constituted the main research area due to its determinant conditions when flood waves occur in the river; the availability of data, and its particularly challenging hydrological characteristics. The hydrologic analysis of páramo watersheds for flood early warning purposes is one of the main contributions of this research, since páramo hydrology is highly complex and still poorly understood, and modelling efforts are few in number. The gaps in knowledge about these types of watersheds contrasts with their importance, since they are the water source of many cities in the tropical Andes, including Bogotá.

In the next subsections, a short summary of each stage of the research is presented and the conclusions are discussed according to the research question that motivated them.

7.1.1   Regional Flood risk analysis

One of the challenges in mountainous areas is the identification of priorities for flood risk management measures. The answer to the frequent and complex question in decision making about where more efforts should be allocated, is addressed in this part of the dissertation that covers chapters 2 and 3. In the former, debris flow susceptibility in mountainous peri-urban areas is addressed and in the latter susceptibility is combined with vulnerability to obtain a proxy for risk level that constitutes the indication of priority for flood risk management. This analysis was carried out in 106 watersheds located in mountainous peri-urban areas of Bogotá.

The research questions for this stage of the research related to debris flow susceptibility are: *When little or no historical information is available, how can hazards produced by debris flows, and by clearwater flows be distinguished using geomorphic data? What physical parameters of the watersheds can be used as reliable indicators of the type of flash flood expected, taking into account highly modified watersheds?*

In this research a susceptibility index composed of morphometric and land cover indicators was proposed. The data to obtain these indicators is derived from digital elevation models and satellite images. The susceptibility index was developed under the assumption that watersheds that are prone to debris flows are more dangerous than other flashy watersheds. Morphometric variables and land cover characteristics showed to be factors that can be efficiently used to reflect the potential susceptibility of the watersheds to clearwater floods or debris flows. The combination of these factors resulted in an index that allowed identifying watersheds more prone than others to potentially more dangerous floods. Morphometric variables were analysed through Principal Component Analysis, finding four principal components that account for 85% of the variance in the data. These correspond to size, shape, hypsometry and potential energy that compose a final morphometric indicator. Land cover was found to exert a significant influence on the susceptibility to different types of flash floods in peri-urban areas. The results show that even if morphometric parameters show a high disposition for debris flow, land cover can compensate and reduce susceptibility. On the contrary, if favourable morphometric conditions are present but deterioration of the land cover of the watershed takes place the danger increases.

Results show that the proposed method is useful in applications at the regional scale for preliminary assessment and identification of more detailed studies. The comparison that was carried out to assess the appropriateness of the susceptibility indicator showed that even if it is relative, it provides useful information that can be used for flood risk management purposes, allowing to efficiently identify high susceptibility watersheds in large mountainous areas.

A drawback of this method is that it does not take sediment availability into account, which is a determining factor of debris flow occurrence. This implies that this aspect must be assessed through other techniques; therefore the method is limited to assess potentiality.

The research question for this stage of the research related to identification of susceptibility areas is: *Can a robust method to determine hazard areas be developed when several geomorphical characteristics of a flashy basin are not known, and to which extent can the methods be simplified to allow reliable identification of the hazard areas even with little data?*

Areas susceptible to floods and debris flows were identified using simplified flood plain delineation methods and the Modified Single Flow Direction (MSF) model. The simplified flood plain delineation methods that were tested correspond to the multi-resolution valley bottom flatness (MRVBF) algorithm and threshold buffers. The MSF showed the capability to model the downstream extent of the flow, however, deviations in the trajectory of the flow caused by obstructions (bridges in the case of the study area) cannot be represented by the model. In order to obtain a delineation of areas susceptible to floods the results of the three methods were combined. The three methods showed limitations to identify flood prone areas in flat topography, therefore, criteria based on the available information and previous studies was needed to estimate reasonable areas of susceptibility. The applied method allows a rapid identification of susceptible areas leading to a conservative delineation, but unavoidable requires further information to limit the extent of susceptible areas in flat topography. Furthermore, the method is limited to susceptibility assessment not hazard, since intensity of the floods associated to probability of occurrence cannot be estimated, only susceptibility areas. The research question for this stage of the research related to prioritisation of flood risks is: *Can a prioritisation method be developed in areas with little data; so critical watersheds from a flood risk perspective can be identified?*

A method of prioritisation of montane watersheds at regional level was proposed. The prioritisation is carried out through a priority index composed of the susceptibility index and a vulnerability index, and is aimed to identify the watersheds with the highest flood damage potential. The susceptibility index was combined with a vulnerability index for the study area to obtain a flood risk proxy useful for prioritisation. Vulnerability was assessed in terms of indicators with the aim to capture the complex interactions that determine vulnerability (physical, social, economic, cultural and educational aspects).

Vulnerability was conceptualized to consist of exposed elements that are intrinsically characterized by physical susceptibility, fragility of the socio economic system and lack of resilience to cope and recover. The susceptibility areas previously obtained from simplified procedures where used as a mask, where exposed elements were identified and where vulnerability variables were extracted.

Principal Component analysis was used to reduce the dimensionality of the variables identified as explanatory of vulnerability and to construct intermediate indicators. A socio-economic fragility indicator, a lack of resilience and coping capacity indicator and a physical exposure indicator were obtained as intermediate indicators. The vulnerability indicator was constructed as the combination of the three intermediate indicators.

In order to combine the vulnerability and susceptibility to derive a level of risk, a classification matrix was used. This was obtained by applying the following steps: (i) from the damage data available for 14 watersheds, these were classified according to damage into high, medium and low; (ii) all possible 3x3 matrices where cells can take high, medium and low levels of priority were defined; (iii) priority was obtained by applying all possible 3x3 matrices and using as input the vulnerability and susceptibility classification for the 14 watersheds, where damage records are available; (iv) the proportion correct of a contingency table comparing priority and damage category for the 14 watersheds was obtained for each possible 3x3 combination matrix; (v) the matrix that produce the highest proportion correct was used for the prioritisation of the whole study area.

The sensitivity analysis of the vulnerability indicator and the prioritisation indicator showed that the method is robust mainly for watersheds with indicator values out of the intermediate ranges where some category changes can occur in a limited amount of watersheds. However, robustness depends on the on the use of three categories of vulnerability, if more categories are introduced, changes in the applied subjective criteria of the analysis may produce significant shifts of category for most watersheds.

The proposed method is flexible to availability of data, which is and advantage for assessments in mountainous developing cities and when the evolution in time of vulnerability variables is taken into account. Further improvements of the method include: (i) the use of smaller units of analysis; (ii) Improvement of physical exposure indicators incorporating type of structures and economic losses; and (iii) incorporation of more detailed information about risk perception and flood early warning.

The prioritisation analysis showed that the areas located in the south of the city are the most critical. These correspond to watersheds with high socio-economic fragility and exposure, where susceptibility is enhanced by poor land cover conditions. This is, where susceptibility to more destructive floods combined with vulnerability conditions can lead to significant impacts. These areas are addressed further in chapter 5, since they are part of the basin chosen for development of flood early warning.

7.1.2 Hydrometeorological inputs

Precipitation and evapotranspiration are the forcing input data to the hydrological models used in this research. Precipitation is addressed in chapter 4 through spatial interpolation. The contribution of this chapter is the proposal and assessment of a robust procedure for real-time interpolation of point measurements of daily precipitation. The performance of Ordinary Kriging and regression Kriging using individual and pooled variograms was assessed in order to identify the potential of simplified interpolation procedures with averaged variograms; and addressing the challenge of defining the contribution of secondary variables in the improvement of the rainfall field. The uncertainty due to choice of interpolation methods on the precipitation volumes was estimated using Gaussian simulation.

The analysis showed the high variability of the storms with concentration of high values of precipitation over small areas, which create difficulty in the determination of the spatial structure using the available rainfall network. This suggests the need to improve the monitoring system in the study area and further research on merging ground, satellite and radar products and production of rainfall fields at sub-daily scales.

The research question for this stage of the research related to identification of susceptibility areas is: *What secondary variables apart from precipitation can be incorporated into the rainfall model to improve the interpolation of precipitation at different time scales?*

A stepwise regression between precipitation and secondary variables, applied to all the datasets individually and afterwards to the standardized precipitation, showed that the east and north coordinates, are the most important secondary variables to improve precipitation estimates at daily scale. However, the choice between incorporating or not the secondary variables in the rainfall field interpolation, which means the choice between Regression Kriging and Ordinary Kriging, can be based on a comparison of the importance of the correlation of precipitation with secondary variables, when compared to the autocorrelation of the data. The use of Regression Kriging with individual residual variograms increases de percentage of variability explained (PVE) of the spatial variability of the daily precipitation in most datasets. Using smoothed secondary variables; the results show an improvement over those found with Ordinary Kriging in most cases. However, the amount of improvement is shown to depend on the relationship between the PVE of Ordinary Kriging and the adjusted R2 of the correlation of precipitation and the smoothed secondary variables. Only when the adjusted R2 is significantly higher than the PEV of OK will a significant improvement be obtained, and where this is not the case considering secondary variables may even be detrimental. Due to this, no interpolator based on KED can be said to be better than OK in all cases. Therefore, the use of secondary variables and the inherent complexity in the procedure should only be considered when the performance or ordinary Kriging is poor in comparison with the correlation of precipitation and secondary variables.

Since the results show that Ordinary Kriging using a climatological variogram as well as Regression Kriging based on an average residual variogram provide robust techniques to obtain rainfall fields in real-time operation for flood early warning purposes, these methods were used in the pre-processing of input precipitation for the hydrological models constructed for the study area. Rainfall fields in the study area are required at sub-daily resolution due to the rapid hydrologic response of the watersheds; however, fitting the variogram for time steps smaller than 24 hours represents the difficulty of considerable scatter for sparse rain gauges with high resolution in the study area. Therefore, the procedures were tested with the daily variograms under the assumption that a proper spatial structure can be provided by the daily data at smaller time steps. The analysis of the hourly rainfall fields showed that Kriging with external drift resulted in unrealistic intensities for the study area in most storm periods (¿100 mm/hr), therefore this interpolation method was not considered further. This result may be due to the scale dependency of the relation between precipitation and secondary variables, which limits its applicability to only daily precipitation interpolation. In the case of Ordinary Kriging, runoff coefficients in the headwater catchments of the study area, showed unrealistically high values larger than 1, indicating the underestimation of the precipitation volume. This behaviour of Ordinary Kriging is due to the sparse hydrometeorological network that provides sub daily data, and therefore interpolation of values beyond the range of the variogram, resulting in most values equalling the mean of the recorded precipitation.

The results of the interpolation with Ordinary Kriging and Kriging with external drift at sub daily scale, lead to the conclusion that although these methods provide good results with the available daily stations, the results are not satisfactory when using the sparse network available at sub daily scales.

7.1.3   Hydrological models for flood early warning

From the priority analysis (chapters 2 and 3), the lower and middle basin of the Tunjuelo River concentrates the watersheds with the highest priority for implementation of flood risk management measures. Thus, the Tunjuelo basin was chosen for hydrological modelling with flood forecasting purposes. Three models were explored, namely HEC-HMS Soil Moisture Accounting - HECHMSSMA (lumped), TOPMODEL (semi distributed) and TETIS (distributed) with the purpose to identify the most convenient modelling approach according to the characteristics of the study area. In the case of the semi distributed and distributed model, resolution was explored in order to identify the most suitable pixel size to be used. Subsequently, a comparison of precipitation input uncertainty and model performance was carried out in order to identify the importance of these in the modelling results.

The upper area of the Tunjuelo basin corresponds to a páramo, which implies soils with a high water storage capacity and high conductivity. Páramo hydrology is a research topic where major gaps in knowledge exist. Rainfall is characterized by high spatial variability; in addition horizontal precipitation, fog and mist in scarce data environments increase the difficulty of the estimation of forcing data for the models. Likewise, evapotranspiration, albeit recognized to be low, is difficult to estimate due to the limitations of the available methods. The hydrologic behaviour of the vegetation present in the páramo areas constitutes another challenge due the current poor understanding of their role. The high water regulation in páramos is due to the slow hydrologic response of the soils, therefore, one of the challenges and contributions of the modelling chapter (chapter 4) is the analysis based on the performance and representation of processes of a still poorly understood hydrologic environment to provide guidance on the selection of models for flood forecasting.

The research questions for this stage of the research are: *What is the most appropriate modeling approach for a páramo watershed? In the case of distributed and semi-distributed models what grid size should be used for appropriate representation of hydrological processes? What is the importance of input and model uncertainty in the modeling results of a páramo watershed?*

From the three models tested, TOPMODEL appears to be the most realistic model for the páramo, although it is more sensitive to rainfall fields variability. This model is able to reproduce the slow water release from the soil layer over the rock stratum that is one of the main characteristics of the páramo soil. The signatures obtained from the flow duration curves show that this is the model that more closely reproduces all ranges of discharge in the three páramo sub-basins. Besides providing more reliability, TOPMODEL demands low computational resources and short run times. These aspects support that TOPMODEL is the preferred choice from a flood early warning perspective.

The use of varying pixel sizes in the semi-distributed and distributed model, showed that a first and determinant criteria for upper limits in pixel size is the ability of the grid to appropriately reproduce the drainage characteristics of a basin. Furthermore, variations in the pixel size are compensated by selected parameters in each model, in order to reach approximately the same performance for all grid sizes. In the case of TOPMODEL the compensation is achieved though variations in the transmisitivy, for TETIS the compensation is manly achieved through variations in the lateral conductivity of the soil. Despite the compensation of parameters, an optimum grid size could be identified in the TETIS and TOPMODEL through the use of the flow duration curve signatures, through which the slight variations in representation of processes could be identified according to pixel size. These optimum grid sizes are 500 meters for TETIS and 25 meters for TOPMODEL.

The comparison of discharge ensembles showed that TOPMODEL is highly sensitive to variations in input precipitation. Increases in precipitation cause a significant increase in the peak discharges. TETIS and HECHMSSMA are less sensitive, showing similarities in performance and behaviour, but without the ability to reproduce the slow water release of the low flow portion of the flow duration curve.

7.2 Added value of the numerical weather prediction model WRF in the flood forecasting system

Numerical weather prediction (NWP) models are crucial to increase lead times in flood forecasting. However, they require a post processing process to correct biases and to produce probabilistic inputs for the hydrological models. Despite their significant advances, the use NWP precipitation forecasts is challenging for flood forecasting in tropical mountainous watersheds due to biases, coarse scale that limits the representation of convection, uncertainty in initial conditions, lack of monitoring data and complex meteorological conditions. In Chapter 5, the WRF model is used to drive the TOPMODEL obtained from the modelling stage of the project. The contribution of this chapter is the assessment of the added value and potential of the WRF model in a flood forecasting system in a páramo area. 107 days with significant storms were chosen to produce forecasts at 00:00 GTM, 6:00 GTM, 12:00 GTM and 18:00 GTM with the WRF model under the current settings used by the meteorological agency of Colombia (IDEAM). These consider three domains with spatial resolutions of 15, 5 and 1.67 km. The observed precipitation originated from the available tipping bucket rainfall gauges. These were used to produce rainfall fields through inverse distance weighing. Forecasts generated under four strategies were used to drive the hydrological model: a) Zero rainfall forecasts; b) raw forecasts from the WRF model; c) bias corrected WRF forecasts through Distribution-Based Scaling - DBS; d) and precipitation forecast ensembles obtained from the post processing of the WRF model. The ensembles were obtained from a two-stage model based on a probit model and quantile regression. The quantile regression was carried out considering two options: normal quantile transformed data and raw data. In the case of deterministic forecasts the Mean Absolute Error (MAE), the Mean Squared Error (MSE), the Mean Error (ME) and the skill score based on the MSE were used for verification and in the case of probabilistic forecasts the ROC (Relative Operating Characteristic) diagram, rank histograms and the continuous ranked probability skill score (CRPSS). For the CRPSS the discharge obtained from forecast precipitation equal to zero was used as reference.

The research questions for this stage of the research are: *What possible improvement can bias correction procedures provide? What is the added value of an NWP model in a flood forecasting system in a páramo area?*

The improvement obtained from the two bias correction methods used in this analysis are summarized as follows:

- Bias correction through Distribution-Based Scaling – DBS: This method significantly reduced the differences in rainfall frequency between observations and forecasts. However, a reduction of performance was observed when intensive precipitation, that is not forecast by the WRF is included in the data. Similar behaviour was observed with the mean of the ensembles with and without normal quantile transformation, albeit with less impact on the performance in comparison with DBS. This undesired behaviour is the result of the inability of the WRF model to adequately represent convective precipitation, which cannot be corrected through simple post-processing.

- Ensembles: Despite the limitations in the WRF forecasts, the ensemble mean outperforms the DBS bias correction and seems to be less sensitive to the presence of highly intense precipitation that is not forecast by the WRF model. No precipitation forecast used in this analysis showed added value when compared to climatology. However, the reduction of biases obtained from the ensembles show potential of this method and model to provide usable precipitation forecasts.

Results show that the streamflow forecasts obtained from a hydrological model driven by post-processed WRF precipitation add value to the flood early warning system when compared to zero precipitation forecasts. The WRF model for the study area provides forecasts that overpredict precipitation and that tend to fail to forecast high intensive precipitation. This behaviour may be due to parametrization deficiencies, errors in boundary conditions and poor representation of topography.

The probit model, used to forecast the precipitation occurrence based on the WRF forecast, showed that the WRF model has some skill at short lead times (up to 5 hours) in comparison to both zero precipitation forecasts and random forecasts, although that skill is limited.

Increases in precipitation biases are reflected in the discharge forecasts. However, discharge forecasts generated with ensembles exhibit higher skill than deterministic forecasts (based on raw, bias corrected and zero precipitation). Furthermore, the quality of these forecasts is better than what could be obtained using zero precipitation as input to the hydrological model.

The potentially usable lead times for forecasts obtained from the WRF model found in this analysis (5-6 hours) are in the range of lead times found in other studies with other NWP models. Despite the fact that the added value of the WRF model forecasts is modest, this shows promise for increasing forecast skill in areas of high meteorological and topographic complexity and the possibility of improvement.

7.3   Recommendations

The following recommendations for further research are suggested.

- The use of indicators showed a high potential not only for assessment of susceptibility to floods but also for risk assessment. The limitations that were identified in this part of the research provide several topics for further research:

  - Improvement of the susceptibility index through incorporation of information on sediment availability. New indicators could be explored using remote sensing data, geology, landslide susceptibility assessments and fieldwork.

  - Further improvement of the priority index include: (i) the use of smaller units of analysis; (ii) Improvement of physical exposure indicators incorporating type of structures and economic losses; and (iii) incorporation of more detailed information about risk perception and flood early warning. Regarding topic iii, insight in the effectiveness of flood early warning in the risk knowledge of individuals would help not only to improve the priority index but to better understand the benefits, limitations and potential improvement of flood early warning systems. These topics imply the increase of availability of data at more detailed scales; therefore innovative methods for collection of data in large areas should be explored.

- Simplified and flexible methods such as the priority index presented in this research could be used to gain insight into the drivers of risk and also to monitor their dynamics. Implementation of the indicator system would help to inform regional risk management decisions. This implies the challenge of updated well-structured databases not only of relevant socio-economic statistics but also of flood events. Further research could be focused on methods to organize, implement and update databases in an efficient way and to identify the variables that influence the most the dynamics of risk.

- The estimation of a rainfall field as input to hydrological models showed to be one of the most challenging topics of this research. Sparse rainfall networks in mountainous environments produce rainfall fields with an associated high uncertainty. A relevant research need is the use of multiple sources of rainfall data (satellite, radar and rainfall gauges) to improve the rainfall field. Páramo areas would require also the improvement in the estimation of water inputs originated from fog events and advancing the understanding of the role of vegetation in the interception and evapotranspiration processes.

- There is a need for more detailed evaluation of the WRF model in this study area. The use of satellite and soon to be available radar data may improve performance. Furthermore, other convective and microphysics schemes should be assessed to identify the most suitable parameterization.

- Further research in the production of precipitation forecasts in tropical mountainous areas is a key topic that requires a joint effort from both the research institutions and the governmental agencies. Advances in meteorological modelling, nowcasting and monitoring improvement are an urgent need. However this advances require a process that is probably not at the same pace, as the risk management needs. Therefore, a coordinated and inclusive work could provide a tool to overcome this issue. Community based approaches could provide alternatives when short lead times limit centralized operation and an added value represented in the shared responsibility of the risk process through the effective involvement of communities.

Bibliography

Agnew, M. and Palutikof, J. (1999). Gis-based construction of baseline climatologies for the mediterranean using terrain variables. *Climate Research*, 14:115 – 127.

Akbas, S., Blahut, J., and Sterlacchini, S. (2009). Critical assessment of existing physical vulnerability estimation approaches for debris flows. In *Landslide processes: from geomorphological mapping to dynamic modelling*, pages 229–233.

Al-Rawas, G. and Valeo, C. (2010). Relationship between wadi drainage characteristics and peak-flood flows in arid northern Oman. *Hydrological Sciences Journal*, 55(3):377–393.

Albano, R., Sole, A., Adamowski, J., and Mancusi, L. (2014). A GIS-based model to estimate flood consequences and the degree of accessibility and operability of strategic emergency response structures in urban areas. *Natural Hazards and Earth System Science*, 14:2847–2865.

Albert, J. (2009). *Bayesian Computation with R*. Springer-Verlag New York, New York, second edition.

Alfieri, L., Pappenberger, F., Wetterhall, F., Haiden, T., Richardson, D., and Salamon, P. (2014). Evaluation of ensemble streamflow predictions in Europe. *Journal of Hydrology*, 517:913–922.

Allen, R. G., Pereira, L. S., Raes, D., and Smith, M. (2006). *Evapotranspiración del cultivo. Guías para la determinación de los requerimientos de agua de los cultivos*. Organización de las Naciones Unidas para la Agricultura y la Alimentación, Roma, Italy.

Antonucci, A., Salvetti, A., and Zaffalon, M. (2007). Credal networks for hazard assessment of debris flows. In Kropp, J. and Scheffran, J., editors, *Advanced Methods for Decision Making and Risk Management in Sustainability*, chapter 10, pages 237–256. Nova Science, New York.

Arango, C. and Ruiz, J. F. (2011). Implementación del modelo WRF para la sabana de Bogotá. Technical report, Instituto de Hidrología, Meteorología y Estudios Ambientales, Bogotá.

Arnaud, P., Bouvier, C., Cisneros, L., and Dominguez, R. (2002). Influence of rainfall spatial variability on flood prediction. *Journal of Hydrology*, 260(1-4):216–230.

Aulitzky, H. (1982). Preliminary two-fold classification of torrents. *Mitteilungen der Forstlichen Bundesversuchsanstalt*, 144:243–256.

Baker, V. R. (1976). Hydrogeomorphic methods for the regional evaluation of flood hazards. *Environmental Geology*, 1(5):261–281.

Balica, S. F., Wright, N. G., and van der Meulen, F. (2012). A flood vulnerability index for coastal cities and its use in assessing climate change impacts. *Natural Hazards*, 64(1):73–105.

Bargaoui, Z. K. and Chebbi, A. (2009). Comparison of two kriging interpolation methods applied to spatiotemporal rainfall. *Journal of Hydrology*, 365(1-2):56 – 73.

Barrenechea, J., Gentile, E., González, S., and Natenson, C. (2000). Una propuesta metodológica para el estudio de la vulnerabilidad social en el marco de la teoría social del riesgo. In Facultad de Ciencias Sociales, U., editor, *IV Jornadas de Sociología*, pages 1–13, Buenos Aires. Facultad de Ciencias Sociales, UBA.

Barroca, B., Bernardara, P., Mouchel, J. M., and Hubert, G. (2006). Indicators for identification of urban flooding vulnerability. *Natural Hazards and Earth System Science*, 6(4):553–561.

Basher, R. (2006). Global early warning systems for natural hazards: systematic and people-centred. *Philosophical transactions. Series A, Mathematical, physical, and engineering sciences*, 364(1845):2167–82.

Baumann, V. and Wick, E. (2011). Debris flow susceptibility mapping at a regional scale along the National Road N7, Argentina. In *2011 Pan-Am CGS Geotechnical Conference*. Accessed: 2013-07-15.

Beltrán, J. (2008). Crecimiento Urbano, Pobreza y Medio Ambiente en Bogotá: Los Efectors Soci Ambientales en Tres Humedales. In *CII Seminario Nacional de Investigación Urbano Regional*, pages 1–13, Medellín - Colombia.

Bernal, G., Rosero, M., Cadena, M., Montealegre, J., and Sanabria, F. (2007). Estudio de la Caracterización Climática de Bogotá y cuenca alta del Río Tunjuelo. Technical report, Instituto de Hidrología, Meteorología y Estudios Ambientales IDEAM - Fondo de Prevención y Atención de Emergencias FOPAE, Bogotá.

Berne, A., Delrieu, G., Creutin, J.-D., and Obled, C. (2004). Temporal and spatial resolution of rainfall measurements required for urban hydrology. *Journal of Hydrology*, 299(3-4):166–179.

Bertrand, M., Liébault, F., and Piégay, H. (2013). Debris-flow susceptibility of upland catchments. *Natural Hazards*, 67(2):497–511.

Beven, K. (1997). TOPMODEL: A Critique. *Hydrological Processes*, 11(December 1996):1069–1085.

Beven, K. (2012). Hydrological Similarity and Distribution Function Rainfall-Runoff models. In *Rainfall-Runoff Modelling*, pages 397–448. John Wiley & Sons, Ltd.

Beven, K. and Binley, A. (1992). The future of distributed models: model calibration and uncertainty prediction. *Hydrological Processes*, 6:279–298.

Beven, K. J. and Kirkby, M. J. (1979). A physically based, variable contributing area model of basin hydrology / Un modèle à base physique de zone d'appel variable de l'hydrologie du bassin versant. *Hydrological Sciences Bulletin*, 24(March 2015):43–69.

Birkmann, J. (2006). *Measuring vulnerability to natural hazards : towards disaster resilient societies*. United Nations University Press, New York, USA, second edition.

Birkmann, J., Cardona, O. D., Carreño, M. L., Barbat, A. H., Pelling, M., Schneiderbauer, S., Kienberger, S., Keiler, M., Alexander, D., Zeil, P., and Welle, T. (2013). Framing vulnerability, risk and societal responses: the MOVE framework. *Natural Hazards*, 67(2):193–211.

Birkmann, J., Cardona, O. D., Liliana Carreno, M., Barbat, A. H., Pelling, M., Schneiderbauer, S., Kienberger, S., Keiler, M., Alexander, D. E., Zeil, P., and Welle, T. (2014). Theoretical and Conceptual Framework for the Assessment of Vulnerability to Natural Hazards and Climate Change in Europe. In Birkmann, J., Kienberger, S., and Alexander, D., editors, *Assessment of Vulnerability to Natural Hazards: A European Perspective*, chapter 1, pages 1–19. Elsevier, Waltham USA.

Blahut, J., Horton, P., Sterlacchini, S., and Jaboyedoff, M. (2010). Debris flow hazard modelling on medium scale: Valtellina di Tirano, Italy. *Natural Hazards and Earth System Science*, 10(11):2379–2390.

Bogner, K., Pappenberger, F., and Cloke, H. L. (2012). Technical Note: The normal quantile transformation and its application in a flood forecasting system. *Hydrology and Earth System Sciences*, 16(2006):1085–1094.

Bonnet-Staub, I. (2000). Debris Flow Hazard Assessment And Mapping; Application To A Case Study In The French Alps. In *ISRM International Symposium*, page 25, Melbourne, Australia. International Society for Rock Mechanics.

Bremnes, J. B. r. (2004). Probabilistic Forecasts of Precipitation in Terms of Quantiles Using NWP Model Output. *Monthly Weather Review*, 132:338–347.

Bring, J. (1994). How to standardize regression coefficients. *The American Statistician*, 48(3):209–213.

Bruijnzeel, L. (2001). Hydrology of tropical montane cloud forests: a reassessment. *Land use and water resources research*, 1(2):1–18.

Bruneau, P., Gascuelodoux, C., Robin, P., Merot, P., and Beven, K. (1995). Sensitivity to space and time resolution of a hydrological model using digital elevation data. *Hydrological Processes*, 9(1):69–81.

Buendía, J. A. T. (2013). *Relaciones socioespaciales en los Cerros Orientales : prácticas , valores y formas de apropiación territorial en torno a las quebradas la Vieja y las Delicias en Bogotá*. PhD thesis, Universidad Colegio Mayor Nuestra Señora del Rosario.

Buytaert, W. (2015). Implementation of the hydrological model TOPMODEL in R.

Buytaert, W. and Beven, K. (2011). Models as multiple working hypotheses: hydrological simulation of tropical alpine wetlands. *Hydrological Processes*, 25(11):1784–1799.

Buytaert, W., Célleri, R., De Bièvre, B., Cisneros, F., Wyseure, G., Deckers, J., and Hofstede, R. (2006a). Human impact on the hydrology of the Andean páramos. *Earth-Science Reviews*, 79(1-2):53–72.

Buytaert, W., Célleri, R., De Biévre, B., Deckers, J., and Wyseure, G. (2005a). Modelando el comportamiento hidrológico de microcuencas de páramo en el Sur del Ecuador con TOP MODEL.

Buytaert, W., Celleri, R., Willems, P., Bièvre, B. D., and Wyseure, G. (2006b). Spatial and temporal rainfall variability in mountainous areas: A case study from the south ecuadorian andes. *Journal of Hydrology*, 329(3-4):413–421.

Buytaert, W., Cuesta-Camacho, F., and Tobón, C. (2011). Potential impacts of climate change on the environmental services of humid tropical alpine regions. *Global Ecology and Biogeography*, 20(1):19–33.

Buytaert, W., De Bièvre, B., Wyseure, G., and Deckers, J. (2004). The use of the linear reservoir concept to quantify the impact of changes in land use on the hydrology of catchments in the Andes Use of linear reservoir concept to quantify the impact of changes in land use on the hydrology of catchments in the Andes. *Hydrology and Earth System Sciences*, 8(1):108–114.

Buytaert, W., Iñiguez, V., and Bièvre, B. D. (2007). The effects of afforestation and cultivation on water yield in the Andean páramo. *Forest Ecology and Management*, 251(1-2):22–30.

Buytaert, W., Iñiguez, V., Celleri, R., Bièvre, B. D., Wyseure, G., Deckers, J., and Célleri, R. (2006c). Analysis of the water balance of small páramo catchments in south Ecuador. *Environmental Role of Wetlands in Headwaters*, pages 271–281.

Buytaert, W., Wyseure, G., De Bièvre, B., and Deckers, J. (2005b). The effect of land-use changes on the hydrological behaviour of Histic Andosols in south Ecuador. *Hydrological Processes*, 19(20):3985–3997.

Calvetti, L., José, A., and Filho, P. (2014). Ensemble Hydrometeorological Forecasts Using WRF Hourly QPF and TopModel for a Middle Watershed. *Advances in Meteorology*, 2014(Volume 2014):12.

Cardona, O. (2003). The need for rethinking the concepts of vulnerability and risk from a holistic perspective: a necessary review and criticism for effective risk management. In Bankoff, G., Frerks, G., and Hilhorst, D., editors, *Mapping Vulnerability: Disasters, Development and People*, chapter 3, pages 37–51. Earthscan Publishers, London.

Cardona, O. D. (2001). *Estimación Holística del Riesgo Sísmico utilizando Sistemas Dinámicos Complejos*. PhD thesis, Universidad Politécnica de Cataluña.

Cardona, O. D., Van Aalst, M. K., Birkmann, J., Fordham, M., McGregor, G., Perez, R., Pulwarty, R., Schipper, L., and Tan Sinh, B. (2012). Determinants of Risk : Exposure and Vulnerability. In Field, C., Barros, V., Socker, T., Qin, D., Dokken, D., Ebi, K., Mastrandrea, M., Mach, K., Plattner, G., Allen, S., Tignor, M., and Midgley, P., editors, *Managing the Risks of Extreme Events and Disasters to Advance Climate Change Adaptation. A Special Report of Working Groups I and II of the Intergovernmental Panel on Climate Change (IPCC).*, chapter 2, pages 65–108. Cambridge University Press, Cambridge, UK, and New York, NY, USA.

Carrera-Hernández, J. and Gaskin, S. (2007). Spatio temporal analysis of daily precipitation and temperature in the basin of mexico. *Journal of Hydrology*, 336:231–249.

Carroll, J. B. (1953). An analytical solution for approximating simple structure in factor analysis. *Psychometrika*, 18(1):23–38.

Carroll, J. B. (1957). Biquartimin Criterion for Rotation to Oblique Simple Structure in Factor Analysis. *Science*, 126(3283):1114–1115.

Cattell, R. B. (1966). The scree test for the numbers of factors. *Multivariate Behavioral Research*, 1(2):245–276.

Célleri, R. and Feyen, J. (2009). The Hydrology of Tropical Andean Ecosystems: Importance, Knowledge Status, and Perspectives. *Mountain Research and Development*, 29(4):350–355.

Chaparro, O. I. (2005). *Evaluación de riesgo por flujos de lodo en la quebrada La Chapa, Municipios de Tasco and Socha (Boyacá)*. PhD thesis, Universidad Nacional de Colombia, Bogotá.

Chen, C.-Y. and Yu, F.-C. (2011). Morphometric analysis of debris flows and their source areas using GIS. *Geomorphology*, 129(3-4):387–397.

Chen, J., Brissette, F. P., Chaumont, D., and Braun, M. (2013). Finding appropriate bias correction methods in downscaling precipitation for hydrologic impact studies over North America. *Water Resources Research*, 49(7):4187–4205.

Chen, T.-c., Wang, H.-y., and Wang, S.-m. (2010). Representative slope index of debris flow streams in Taiwan. In *Interpraevent 2010*, pages 91–99, Taipei.

Chen, Y., Barrett, D., Liu, R., and Gao, L. (2014). A spatial framework for regional-scale flooding risk assessment. In Ames, D., Quinn, N., and Rizzoli, A., editors, *7th International Congress on Environmental Modelling and Software*, pages 1777–1783, San Diego, California, USA.

Chiles, J. P. and Delfiner, P. (1999). *Geostatistics: modelling spatial uncertainty.* Series in probability and statistics applied probability and statistics section. Wiley-Interscience, New York.

Cimmery, V. (2010). *User Guide for SAGA (version 2.0.5), Vol. 2*, volume 2.

Clark, M., Gangopadhyay, S., Hay, L., Rajagopalan, B., and Wilby, R. (2004). The Schaake Shuffle: A Method for Reconstructing Space–Time Variability in Forecasted Precipitation and Temperature Fields. *Journal of Hydrometeorology*, 5:243–262.

Clark, M. P. and Slater, A. G. (2006). Probabilistic quantitative precipitation estimation in complex terrain. *Journal of Hydrometeorology*, 7(1):3–22. doi: 10.1175/JHM474.1.

Clark, M. P., Slater, A. G., Rupp, D. E., Woods, R. a., Vrugt, J. a., Gupta, H. V., Wagener, T., and Hay, L. E. (2008). Framework for Understanding Structural Errors (FUSE): A modular framework to diagnose differences between hydrological models. *Water Resources Research*, 44(12):1–14.

Cloke, H. L. and Pappenberger, F. (2009). Ensemble flood forecasting: A review. *Journal of Hydrology*, 375(3-4):613–626.

Cluckie, I. D., Xuan, Y., and Wang, Y. (2006). Uncertainty analysis of hydrological ensemble forecasts in a distributed model utilising short-range rainfall prediction. *Hydrology and Earth System Sciences Discussions*, 3(5):3211–3237.

Cohen, S., Willgoose, G., and Hancock, G. (2008). A methodology for calculating the spatial distribution of the area-slope equation and the hypsometric integral within a catchment. *Journal of Geophysical Research*, 113(F3):1–13.

Colman, B., Cook, K., and Snyder, B. (2013). Numerical weather prediction and weather forecasting in complex terrain. In Chow, F. K., De Wekker, S. F., and Snyder, B. J., editors, *Mountain Weather Research and Forecasting Recent Progress and Current Challenges*, chapter 11. Springer, Dordrecht.

Costa, J. (1988). Rheologic, geomorphic, and sedimentologic differentiation of water floods, hyper-concentrated flows, and debris flows. In Baker, V. R., Kochel, R. C., and Patton, P. C., editors, *Flood Geomorphology*, pages 113–122. Wiley, New York.

Crespo, P., Celleri, R., Buytaert, W., and Feyen, J. (2009). Land use change impacts on the hydrology of wet Andean páramo ecocystems. In Hermann, A. and Schumann, S., editors, *Proceedings of the*

International Workshop on Status and Perspectives of Hydrology in Small Basins, pages 71–76, Goslar-Hahnenklee, Germany. IAHS Press.

Crespo, P., Feyen, J., Buytaert, W., Célleri, R., Frede, H.-G., Ramírez, M., and Breuer, L. (2012). Development of a conceptual model of the hydrologic response of tropical Andean micro-catchments in Southern Ecuador. *Hydrology and Earth System Sciences Discussions*, 9(2):2475–2510.

Crespo, P. J., Feyen, J., Buytaert, W., Bücker, A., Breuer, L., Frede, H.-G., and Ramírez, M. (2011). Identifying controls of the rainfall–runoff response of small catchments in the tropical Andes (Ecuador). *Journal of Hydrology*, 407(1-4):164–174.

Crosta, G. B. and Frattini, P. (2004). Controls on modern alluvial fan processes in the central Alps, northern Italy. *Earth Surface Processes and Landforms*, 29(3):267–293.

Cuo, L., Pagano, T. C., and Wang, Q. J. (2011). A Review of Quantitative Precipitation Forecasts and Their Use in Short- to Medium-Range Streamflow Forecasting. *Journal of Hydrometeorology*, 12(5):713–728.

Cutter, S. L., Barnes, L., Berry, M., Burton, C., Evans, E., Tate, E., and Webb, J. (2008). A place-based model for understanding community resilience to natural disasters. *Global Environmental Change*, 18(4):598–606.

Cutter, S. L., Boruff, B. J., and Shirley, W. L. (2003). Social vulnerability to environmental hazards. *Social Science Quarterly*, 84(2):242–261.

Daly, C., Neilson, R. P., and Phillips, D. L. (1993). A statistical-topographic model for mapping climatological precipitation over mountainous terrain. *Journal of Applied Meteorology*, 33(2):140–158.

Daza, M. C., Florez, F. H., and Triana, F. A. (2014). Efecto del Uso del Suelo en la Capacidad de Almacenamiento Hídrico en el Páramo de Sumapaz - Colombia. *Revista Facultad Nacional de Agronomía*, 67(1):7189–7200.

de Matauco, G. and Ibisate, A. (2004). Análisis morfométrico de la cuenca y de la red de drenaje del río Zadorra y sus afluentes aplicado a la peligrosidad de crecidas. *Boletín de la Asociación de Geógrafos Españoles*, (38):311–329.

De Scally, F., Owens, I., and Louis, J. (2010). Controls on fan depositional processes in the schist ranges of the Southern Alps, New Zealand, and implications for debris-flow hazard assessment. *Geomorphology*, 122(1-2):99–116.

De Scally, F. A. and Owens, I. F. (2004). Morphometric controls and geomorphic responses on fans in the Southern Alps, New Zealand. *Earth Surface Processes and Landforms*, 29(3):311–322.

Deginet, M. D. (2008). *Land surface representation for regional rainfall-runoff modelling, upper Blue Nile basin, Ethiopia*. Master of science, International Institute for Geo-Information Science and Earth Observation.

Demeritt, D., Cloke, H., Pappenberger, F., Thielen, J., Bartholmes, J., and Ramos, M. H. (2007). Ensemble predictions and perceptions of risk, uncertainty, and error in flood forecasting. *Environmental Hazards*, 7(2):115–127.

Demeritt, D., Nobert, Ś., Cloke, H., and Pappenberg, F. (2010). Challenges in communicating and using ensembles in operational flood forecasting. *Meteorological Applications*, 17(2):209–222.

Deraisme, J., Humbert, J., Drogue, G., and Freslon, N. (2001). Geostatistical interpolation of rainfall in mountainous areas. In Monestiez, P., editor, *GeoENV III: Geostatistics for Environmental Applications*, Dordrecht. Kluwer Academic Publishers.

Di, Z., Qingyun, D., Wei, G., Chen, W., Yanjun, G., Jiping, Q., Jianduo, L., Chiyuan, M., Aizhong, Y., and Charles, T. (2014). Assessing WRF model parameter sensitivity: A case study with 5 day summer precipitation forecasting in the Greater Beijing Area. *Geophysical Research Letters*, pages 579–587.

Díaz-granados, M., Céspedes, D., Tamayo, A., Clavijo, W., and Saénz, J. (2002). SIG en el Estudio de Cuencas de Páramo. In Jaramillo, C. A., Uribe, C., Hincapié, F., Rodríguez, J., and Durán, C., editors, *Congreso Mundial de Páramos*, pages 698–704, Bogotá. Ministerio del Medio Ambiente.

Díaz-Granados, M., González, J., and López, T. (2005). Páramos: Hidrosistemas Sensibles. *Revista de Ingeniería Universidad de Los Andes*.

Diodato, N. (2005). The influence of topographic co-variables on the spatial variability of precipitation over small regions of complex terrain. *International Journal of Climatology*, 25(3):351–363.

Diodato, N. and Ceccarelli, M. (2005). Interpolation processes using multivariate geostatistics for mapping of climatological precipitation mean in the sannio mountains (southern italy). *Earth Surface Processes and Landforms*, 30(3):259–268.

DPAE (2003a). Diagnóstico Técnico 1836. Technical report, Direccion de Prevención y Atención de Emergencias de Bogotá, Bogotá.

DPAE (2003b). Diagnóstico Técnico 1891. Technical report, Direccion de Prevención y Atención de Emergencias de Bogotá, Bogotá.

DPAE (2003c). Dianóstico Técnico 1880. Technical report, Dirección de Prevención y Atención de Emergencias de Bogotá, Bogotá.

DPAE (2005). Diagnóstico Técnico 2414. Technical report, Direccion de Prevención y Atención de Emergencias de Bogotá, Bogotá.

Duan, Q.-Y. Y., Sorooshian, S., and Gupta, V. (1992). Effective and efficient global optimization for conceptual rainfall-runoff models. *Water Resources Research*, 28:1015–1031.

Efstratiadis, A. and Koutsoyiannis, D. (2010). One decade of multi-objective calibration approaches in hydrological modelling: a review. *Hydrological Sciences Journal*, 55(1):58–78.

Ehret, U., Zehe, E., Wulfmeyer, V., Warrach-Sagi, K., and Liebert, J. (2012). Should we apply bias correction to global and regional climate model data? *Hydrology and Earth System Sciences Discussions*, 9(4):5355–5387.

Esty, D., Srebotnjak, T., Kim, C., Levy, M., Sherbinin, A., and Anderson, B. (2006). Pilot 2006 Environmental Performance Index. Technical report, Yale Center for Environmental Law & Policy, New Haven, United States of America.

Euser, T., Winsemius, H. C., Hrachowitz, M., Fenicia, F., Uhlenbrook, S., and Savenije, H. H. G. (2013). A framework to assess the realism of model structures using hydrological signatures. *Hydrology and Earth System Sciences*, 17(5):1893–1912.

Fan, F. M., Collischonn, W., Meller, A., and Botelho, L. C. M. (2014). Ensemble streamflow forecasting experiments in a tropical basin: The São Francisco river case study. *Journal of Hydrology*, 519:2906–2919.

Faulkner, H., Parker, D., Green, C., and Beven, K. (2007). Developing a translational discourse to communicate uncertainty in flood risk between science and the practitioner. *Ambio*, 36(8):692–703.

Fekete, a. (2009). Validation of a social vulnerability index in context to river-floods in Germany. *Natural Hazards and Earth System Science*, 9(2):393–403.

Fenicia, F., Kavetski, D., and Savenije, H. H. G. (2011). Elements of a flexible approach for conceptual hydrological modeling: 1. Motivation and theoretical development. *Water Resources Research*, 47(November):1–13.

Fenicia, F., Savenije, H. H. G., Matgen, P., and Pfister, L. (2007). A comparison of alternative multiobjective calibration strategies for hydrological modeling. *Water Resources Research*, 43:1–16.

Fenicia, F., Savenije, H. H. G., Matgen, P., and Pfister, L. (2008). Understanding catchment behavior through stepwise model concept improvement. *Water Resources Research*, 44(1):1–13.

Fleming, M. and Doan, J. (2013). HEC-GeoHMS Geospatial Hydrologic Modeling Estension. Technical report, US Army Corps of Engineers, Davis.

Fleming, M. and Neary, V. (2004). Continuous hydrologic modeling study with the hydrologic modeling system. *Journal of Hydrologic Engineering*, 9(3).

FLO-2D Software, I. (2006). *FLO-2D Users manual (Version 2006.01)*. Arizona, USA.

Frances, F. (2012). Descripción del modelo conceptual distribuido de simulacion hidrológica TETIS v.8. Technical report, Universidad Politécnica de Valencia, Valencia, España.

Franchini, M., Wendling, J., Obled, C., and Todini, E. (1996). Physical interpretation and sensitivity analysis of the TOPMODEL. *Journal of Hydrology*, 175:293–338.

Franz, K. J. and Hogue, T. S. (2011). Evaluating uncertainty estimates in hydrologic models: borrowing measures from the forecast verification community. *Hydrology and Earth System Sciences*, 15(11):3367–3382.

Frei, C. M. (2012). *Probabilistic Forecasts of Precipitation Using Quantiles*. PhD thesis, Heidelberg University.

Fuchs, S. (2009). Susceptibility versus resilience to mountain hazards in Austria - paradigms of vulnerability revisited. *Natural Hazards and Earth System Science*, 9(2):337–352.

Fuchs, S., Heiss, K., and Hübl, J. (2007). Towards an empirical vulnerability function for use in debris flow risk assessment. *Natural Hazards and Earth System Science*, 7(5):495–506.

Fuchs, S. and Holub, M. (2012). Reducing Physical Vulnerability to Mountain Hazards. In *12th Congress INTERPRAEVENT 2012*, pages 675–686, Grenoble / France.

Fuchs, S., Tsao, T.-C., and Keiler, M. (2012). Quantitative Vulnerability Functions for use in Mountain Hazard Risk Management. In *12th Congress INTERPRAEVENT 2012*, pages 885–896, Grenoble / France.

Gallant, J. and Dowling, T. (2003). A multiresolution index of valley bottom flatness for mapping depositional areas. *Water Resources Research*, 39(12):1347–1360.

Garcia, M., Peters-Lidard, C. D., and Goodrich, D. C. (2008). Spatial interpolation of precipitation in a dense gauge network for monsoon storm events in the southwestern united states. *Water Resources Research*, 44(5):1 – 14.

Giraud, R. (2005). *Guidelines for the geologic evaluation of debris-flow hazards on alluvial fans in Utah*. Utah Geological Survey, Utah. Accessed: 2013-07-15.

Goovaerts, P. (1997). *Geostatistics for Natural Resources Evaluation*. Oxford University Press, New York.

Goovaerts, P. (2000). Geostatistical approaches for incorporating elevation into the spatial interpolation of rainfall. *Journal of Hydrology*, 228:113–129.

Gray, D. M. (1961). Interrelationships of Watershed Characteristics. *Journal of Geophysical Research*, 66(4):1215–1223.

Gregory, K. and Walling, D. (1968). The variation of drainage density within a catchment. *Hydrological Sciences Journal*, 13(2):61–68.

Greiving, Stefan (2006). Multi-risk assessment of Europe regions. In Birkmann, J., editor, *Measuring Vulnerability to Natural Hazards: Toward Disaster Resilient Societies*, chapter 11, pages 210–226. United Nations University, New York, USA.

Griffiths, P. G., Webb, R. H., and Melis, T. S. (2004). Frequency and initiation of debris flows in Grand Canyon, Arizona. *J. Geophys. Res.*, 109(F4):F04002.

Grimes, D. I. F., Pardo-Iguzquiza, E., and Bonifacio, R. (1999). Optimal areal rainfall estimation using raingauges and satellite data. *Journal of Hydrology*, 222:93–108.

Grimes, D. I. F. and Pardo-Igúzquiza, E. (2010). Geostatistical analysis of rainfall. *Geographical Analysis*, 42(2):136–160.

Gruber, S., Huggel, C., and Pike, R. (2009). Modelling mass movements and landslide susceptibility. In Hengl, T. and Reuter, H., editors, *Geomorphometry: concepts, software, applications*, chapter 23, pages 527–550. Elsevier, Amsterdam, first edition.

Guan, H., Wilson, J. L., and Makhnin, O. (2005). Geostatistical mapping of mountain precipitation incorporating autosearched effects of terrain and climatic characteristics. *Journal of Hydrometeorology*, 6(6):1018–1031.

Güntner, A., Uhlenbrook, S., Seibert, J., and Leibundgut, C. (1999). Multi-criterial validation of TOPMODEL in a mountainous catchment. *Hydrological Processes*, 13(11):1603–1620.

Gupta, H. V., Kling, H., Yilmaz, K. K., and Martinez, G. F. (2009). Decomposition of the mean squared error and NSE performance criteria: Implications for improving hydrological modelling. *Journal of Hydrology*, 377(1-2):80–91.

Haberlandt, U. (2007). Geostatistical interpolation of hourly precipitation from rain gauges and radar for a large-scale extreme rainfall event. *Journal of Hydrology*, 332(1-2):144–157.

Habets, F., LeMoigne, P., and Noilhan, J. (2004). On the utility of operational precipitation forecasts to served as input for streamflow forecasting. *Journal of Hydrology*, 293(1-4):270–288.

Hack, J. (1957). Studies of longitudinal stream profiles in Virginia and Maryland. *US Geol. Survey Prof. Paper*. Accessed: 2013-07-15.

Haerter, J. O., Hagemann, S., Moseley, C., and Piani, C. (2011). Climate model bias correction and the role of timescales. *Hydrology and Earth System Sciences*, 15:1065–1079.

Hancock, G. R. (2005). The use of digital elevation models in the identification and characterization of catchments over different grid scales. *Hydrological Processes*, 19(9):1727–1749.

Hao, L. and Naiman, D. (2007). *Quantile Regression*. SAGE Publications, Thousand Oaks, California.

Harlin, J. (1978). Statistical moments of the hypsometric curve and its density function. *Mathematical Geology*, 10(1):59–72.

Harlin, J. (1984). Watershed morphometry and time to hydrograph peak. *Journal of Hydrology*, 67(1-4):141–154.

Harris, C. W. and Kaiser, H. F. (1964). Oblique factor analytic solutions by orthogonal transformations. *Psychometrika*, 29(4):347–362.

Hay, L., Viger, R., and McCabe, G. (1998). Precipitation interpolation in mountainous regions using multiple linear regression. In AHS, editor, *Proceedings of the HeadWater'98 Conference*, Meran/Merano, Italy.

Hendrickson, A. and White, P. (1964). Promax: A quick method for rotation to oblique simple structure. *British Journal of Statistical Psychology*, 17(1):65–70.

Hengl, T. (2009). *A practical guide to geostatistical analysis*. EUR 22904 EN. Joint Research Centre, Institute for Environment and Sustainability, Luxembourg.

Hersbach, H. (2000). Decomposition of the Continuous Ranked Probability Score for Ensemble Prediction Systems. *Weather and Forecasting*, 15(5):559–570.

Hevesi, J. A., Istok, J. D., and Flint, A. L. (1992). Precipitation estimation in mountainous terrain using multivariate geostatistics. part i: Structural analysis. *Journal of Applied Meteorology*, 31(7):661–676.

Hofstede, R. G. M., Chilito, E. J. P., and Sandovals, E. M. (1995). Vegetative structure, microclimate, and leaf growth of a páramo tussock grass species, in undisturbed, burned and grazed conditions. *Vegetatio*, 119(1):53–65.

Hollis, G. E. (1975). The effect of urbanization on floods of different recurrence interval. *Water Resources Research*, 11(3):431–435.

Holub, M., Suda, J., and Fuchs, S. (2012). Mountain hazards: Reducing vulnerability by adapted building design. *Environmental Earth Sciences*, 66(7):1853–1870.

Horn, J. L. (1965). A rationale and test for the number of factors in factor analysis. *Psychometrika*, 30(2):179–185.

Horton, P. and Jaboyedoff, M. (2008). Debris flow susceptibility mapping at a regional scale. *4th Canadian Conference on Geohazards,*. Accessed: 2013-07-30.

Hu, K. H., Cui, P., and Zhang, J. Q. (2012). Characteristics of damage to buildings by debris flows on 7 August 2010 in Zhouqu, Western China. *Natural Hazards and Earth System Science*, 12(May 1998):2209–2217.

Huang, X. and Niemann, J. D. (2008). How do streamflow generation mechanisms affect watershed hypsometry? *Earth Surface Processes and Landforms*, 33(5):751–772.

Hufschmidt, G., Crozier, M., and Glade, T. (2005). Evolution of natural risk: research framework and perspectives. *Natural Hazards and Earth System Science*, 5(3):375–387.

Huggel, C., Kääb, A., Haeberli, W., and Krummenacher, B. (2003). Regional-scale GIS-models for assessment of hazards from glacier lake outbursts: evaluation and application in the Swiss Alps. *Natural Hazards and Earth System Science*, 3(6):647–662.

Hurtrez, J., Sol, C., and Lucazeau, F. (1999). Effect of drainage area on hypsometry from an analysis of small scale drainage basins in the Siwalik Hills (Central Nepal). *Earth Surface Processes and Landforms*, 808:799–808.

Hutchinson, M. F. (1998). Interpolation of rainfall data with thin plate smoothing splines - part ii: Analysis of topographic dependence. *Journal of Geographic Information and Decision Analysis*, 2(2,):52 –167.

Hyndman, D. W. and Hyndman, D. W. (2008). *Natural hazards and disasters*. Yolanda Cossio, Belmont USA, 4 edition.

IGAC (2000). Estudio General de Suelos y Zonificacion de Tierras del Departamento de Cundinamarca. Technical report, Instituto Colombiano Agustín Codazzi, Bogotá.

INGETEC (2002). Diseños para la construcción de las obras de control de crecientes de la cuenca del Río Tunjuelo. Technical report, Empresa de Acueducto y Alcantarillado de Bogotá, Bogotá.

IRH (1995). Análisis y caracterización de tormentas en la sabana de bogotá: informe final. Technical report, Empresa de Acueducto y Alcantarillado de Bogotá, Bogotá.

Isaaks, E. H. and Srivastava, R. M. (1989). *Applied Geostatistics*. Oxford University Press, New York.

ISDR (2006). *Developing Early Warning Systems : A Checklist. EWC III Third International Conference on Early Warning from concept to action*. ISDR, Bonn, Germany.

IWR (2011). Flood Risk Management Approaches: As Being Practiced in Japan, Netherlands, United Kingdom and United States. Technical report, United States Army Corps of Engineers, Washington, D.C., USA.

Jackson, L., Kostaschuk, R., and MacDonald, G. (1987). Identification of debris flow hazard on alluvial fans in the Canadian Rocky Mountains. *Geol. Soc. Amer., Rev. eng. Geol.*, 7:115–124.

Jakob, M. (1996). *Morphometric and geotechnical controls of debris flow frequency and magnitude in southwestern British Columbia.* PhD thesis, University of British Columbia.

Jakob, M., Holm, K., Weatherly, H., Liu, S., and Ripley, N. (2013). Debris flood risk assessment for Mosquito Creek, British Columbia, Canada. *Natural Hazards*, 65(3):1653–1681.

Jakob, M., Porter, M., Savigny, K. W., and Yaremko, E. (2004). A geomorphic approach to the design of pipeline crossings of mountain streams. In *Proceedings of IPC 2004: International Pipeline Conference*, pages 1–8, Calgary, Alberta, Canada.

Jakob, M., Stein, D., and Ulmi, M. (2012). Vulnerability of buildings to debris flow impact. *Natural Hazards*, 60:241–261.

Jakob, M. and Weatherly, H. (2005). Debris flow hazard and risk assessment, Jones Creek, Washington. In Hungr, O., Fell, R., Couture, R., and Eberhardl, E., editors, *Landslide risk management. Proceedings*, pages 533–541, London. Taylor & Francis.

Jankov, I., Gallus, W. a., Segal, M., Shaw, B., and Koch, S. E. (2005). The Impact of Different WRF Model Physical Parameterizations and Their Interactions on Warm Season MCS Rainfall. *Weather and Forecasting*, 20(6):1048–1060.

Jha, A. K., Bloch, R., and Lamond, J. (2012). *Cities and Flooding: A Guide to Integrated Urban Flood Risk Management for the 21st Century.* The World Bank, Washington D.C., USA.

JICA (2006). Study on monitoring and early warning systems for landslide and floods in Bogotá and Soacha. Technical report, Japanese Internation Cooperation Agency - JICA, Bogotá.

Johansson, B. and Chen, D. (2003). The influence of wind and topography on precipitation distribution in sweden: statistical analysis and modelling. *International Journal of Climatology*, 23(12):1523–1535. 10.1002/joc.951.

Jolliffe, I. and Stephenson, D. (2003). *Forecast Verification: A Practitioner's Guide in Atmospheric Science.* Wiley, West Sussex, England.

Jolliffe, I. T. (2002). *Principal Component Analysis.* Springer Series in Statistics. Springer-Verlag, New York.

Jonkman, S., Bočkarjova, M., Kok, M., and Bernardini, P. (2008). Integrated hydrodynamic and economic modelling of flood damage in the Netherlands. *Ecological Economics*, 66(1):77–90.

Kaiser, H. F. (1958). The varimax criterion for analytic rotation in factor analysis. *Psychometrika*, 23(3):187–200.

Kaiser, H. F. (1960). The application of electronic computers to factor analysis. *Educational and Psychological Measurement*, 20(1):141–151.

Kappes, M., Papathoma-Köhle, M., and Keiler, M. (2012). Assessing physical vulnerability for multi-hazards using an indicator-based methodology. *Applied Geography*, 32(2):577–590.

Kappes, M. S., Malet, J.-P., Remaître, A., Horton, P., Jaboyedoff, M., and Bell, R. (2011). Assessment of debris-flow susceptibility at medium-scale in the Barcelonnette Basin, France. *Natural Hazards and Earth System Science*, 11(2):627–641.

Kavetski, D. and Fenicia, F. (2011). Elements of a flexible approach for conceptual hydrological modeling: 2. Application and experimental insights. *Water Resources Research*, 47(11):1–19.

Kiers, H. A. L. (1994). Simplimax: Oblique rotation to an optimal target with simple structure. *Psychometrika*, 59(4):567–579.

Kirchner, J. W. (2006). Getting the right answers for the right reasons: Linking measurements, analyses, and models to advance the science of hydrology. *Water Resources Research*, 42(3):1–5.

Kleiber, W., Katz, R. W., and Rajagopalan, B. (2012). Daily spatiotemporal precipitation simulation using latent and transformed Gaussian processes. *Water Resources Research*, 48(January):1–17.

Koenker, R. and Machado, J. a. F. (1999). Goodness of Fit and Related Inference Processes for Quantile Regression. *Journal of the American Statistical Association*, 94(November 2014):1296–1310.

Koks, E., Jongman, B., Husby, T., and Botzen, W. (2015). Combining hazard, exposure and social vulnerability to provide lessons for flood risk management. *Environmental Science & Policy*, 47:42–52.

Koscielny, M., Cojean, R., and Thenevin, I. (2009). Debris flow hazards due to land use change above source areas in torrent catchments . The case study of Les Arcs. *River Basin Management*, 124:161–170.

Kostaschuk, R. (1986). Depositional process and alluvial fan drainage basin morphometric relationships near banff, Alberta, Canada. *Surface Processes and Landforms*, 11:471–484.

Kryza, M., Werner, M., Wałaszek, K., and Dore, A. J. (2013). Application and evaluation of the WRF model for high-resolution forecasting of rainfall - A case study of SW Poland. *Meteorologische Zeitschrift*, 22(5):595–601.

Krzysztofowicz, R. (1999). Bayesian theory of probabilistic forecasting via deterministic hydrologic model. *Water Resources Research*, 35(9):2739–2750.

Kyriakidis, P. C., Kim, J., and Miller, N. L. (2001). Geostatistical mapping of precipitation from rain gauge data using atmospheric and terrain characteristics. *Journal of Applied Meteorology*, 40(11):1855–1877.

Laing, A. and Evans, J.-L. (2010). *Introduction to tropical meteorology*. The Comet Program.

Larsen, M., Wieczorek, G., Eaton, L., and Torres-Sierra, H. (2001). Natural Hazards on Aluvial Fans: The Debris Flow and Flash flood disaster of December 1999, Vargas State, Venezuela. In Sylva, W., editor, *Proceedings of the Sixth Caribbean Islands Water Resources Congress*, pages 1–7, Mayagüez, Puerto Rico.

Lavigne, F. and Suwa, H. (2004). Contrasts between debris flows, hyperconcentrated flows and stream flows at a channel of Mount Semeru, East Java, Indonesia. *Geomorphology*, 61(1-2):41–58.

Lebel, T. and Bastin, G. (1985). Variogram identification by the mean-squared interpolation error method with application to hydrologic fields. *Journal of Hydrology*, 77(1-4):31–56.

Lebel, T. and Bastin, G. (1989). Real-time estimation of areal rainfall with low density telemetered networks. In *Vancouver Workshop*, volume 178, Vancouver. IAHS.

Lebel, T., Bastin, G., Obled, C., and Creutin, J. D. (1987). On the accuracy of areal rainfall estimation: A case study. *Water Resources Research*, 23(11):2123–2134.

Leutbecher, M. and Palmer, T. N. (2008). Ensemble forecasting. *Journal of Computational Physics*, 227(7):3515–3539.

Liu, J., Wang, J., Pan, S., Tang, K., Li, C., and Han, D. (2015). A real-time flood forecasting system with dual updating of the NWP rainfall and the river flow. *Natural Hazards*, 77(2):1161–1182.

Liu, X. and Lei, J. (2003). A method for assessing regional debris flow risk: an application in Zhaotong of Yunnan province (SW China). *Geomorphology*, 52(3-4):181–191.

Liu, Y., Zhou, J., Song, L., Zou, Q., Guo, J., and Wang, Y. (2014). Efficient GIS-based model-driven method for flood risk management and its application in central China. *Natural Hazards and Earth System Science*, 14(2):331–346.

Lo, W.-C., Tsao, T.-C., and Hsu, C.-H. (2012). Building vulnerability to debris flows in Taiwan: a preliminary study. *Natural Hazards*, 64(3):2107–2128.

López López, P., Verkade, J. S., Weerts, a. H., and Solomatine, D. P. (2014). Alternative configurations of quantile regression for estimating predictive uncertainty in water level forecasts for the upper Severn River: A comparison. *Hydrology and Earth System Sciences*, 18(9):3411–3428.

Luino, F., Cirio, C. G., Biddoccu, M., Agangi, A., Giulietto, W., Godone, F., and Nigrelli, G. (2009). Application of a model to the evaluation of flood damage. *GeoInformatica*, 13(3):339–353.

Luino, F., Turconi, L., Petrea, C., and Nigrelli, G. (2012). Uncorrected land-use planning highlighted by flooding: the Alba case study (Piedmont, Italy). *Natural Hazards and Earth System Science*, 12(7):2329–2346.

Luna, B., Blahut, J., Kappes, M., Akbas, S. O., Malet, J. P., Remaître, A., and Jaboyedoff, M. (2014). Methods for Debris Flow Hazard and Risk Assessment. In Van Asch, T., Corominas, J., Greiving, S., Malet, J.-P., and Sterlacchini, S., editors, *Mountain Risks: From Prediction to Management and Governance*, pages 133–177. Springer Netherlands, Dordrecht, The Netherlands.

Luo, W. and Harlin, J. M. (2003). A theoretical travel time based on watershed hypsometry. *JAWRA Journal of the American Water Resources Association*, 39(4):785–792.

Mantilla, R., Mesa, O., and Poveda, G. (2000). Sobre la existencia de la ley de Hack en las cuencas hidrográficas de Colombia. In *XIV Seminario Nacional de Hidráulica e hidrología*. Accessed: 2013-08-08.

Marquinez, J., Lastra, J., and García, P. (2003). Estimation models for precipitation in mountainous regions: the use of gis and multivariate analysis. *Journal of Hydrology*, 270:1–11.

Maskrey, A. (1997). National and Local Capabilities for Early Warning. Technical Report October, United Nations, Lima Peru.

Mazzorana, B., Levaggi, L., Keiler, M., and Fuchs, S. (2012). Towards dynamics in flood risk assessment. *Natural Hazards and Earth System Science*, 12(11):3571–3587.

McMillan, H. K., Clark, M. P., Bowden, W. B., Duncan, M., and Woods, R. a. (2011). Hydrological field data from a modeller's perspective: Part 1. Diagnostic tests for model structure. *Hydrological Processes*, 25(4):511–522.

Mesa, O. J. (1987). On the Main Channel Length-Area Relationship for Channel Networks. *Water Resources*, 23(11):2119–2122.

Molinari, D., Menoni, S., and Ballio, F. (2013). *Flood Early Warning Systems: Knowledge and Tools for Their Critical Assessment*. WIT Press, Southampton, UK.

Montgomery, D. R. and Foufoula-Georgiou, E. (1993). Channel network source representation using digital elevation models. *Water Resources Research*, 29(12):3925–3934.

Mourre, L., Condom, T., Junquas, C., Lebel, T., Sicart, J. E., Figueroa, R., and Cochachin, A. (2015). Spatio-temporal assessment of WRF, TRMM and in situ precipitation data in a tropical mountain environment (Cordillera Blanca, Peru). *Hydrology and Earth System Sciences Discussions*, 12(7):6635–6681.

Muggeo, V. M. R., Sciandra, M., Tomasello, A., and Calvo, S. (2013). Estimating growth charts via nonparametric quantile regression: A practical framework with application in ecology. *Environmental and Ecological Statistics*, 20(4):519–531.

Müller, A., Reiter, J., and Weiland, U. (2011). Assessment of urban vulnerability towards floods using an indicator-based approach – a case study for Santiago de Chile. *Natural Hazards and Earth System Science*, 11(8):2107–2123.

Nardi, F., Vivoni, E. R., and Grimaldi, S. (2006). Investigating a floodplain scaling relation using a hydrogeomorphic delineation method. *Water Resources Research*, 42(9):1–15.

Nardo, M., Saisana, M., Saltelli, A., Tarantola, S., Hoffman, A., and Giovannini, E. (2008). *Handbook on Constructing Composite Indicators*. OECD publishing, Ispra, Italy.

Nash, J. E. and Sutcliffe, J. V. (1970). River flow forecasting through conceptual models part I - A discussion of principles. *Journal of Hydrology*, 10(3):282–290.

Neuhaus, J. O. (1954). The Quartimax Method: An Analytic Approach to Orthogonal Simple Structure. *Brit. J. statist. Psychol.*, 7:81–91.

Ninyerola, M., Pons, X., and Roure, J. M. (2000). A methodological approach of climatological modelling of air temperature and precipitation through gis techniques. *International Journal of Climatology*, 20(14):1823–1841.

Nkwunonwo, U., Whitworth, M., and Baily, B. (2015). Relevance of Social Vulnerability Assessment to Flood Risk Reduction in the Lagos Metropolis of Nigeria. *British Journal of Applied Science & Technology*, 8(4):366–382.

NOAA and COMET (2010). *Flash Flood Early Warning System Reference Guide*. NOAA, Silver Spring, USA.

Ochoa, a., Pineda, L., Crespo, P., and Willems, P. (2014). Evaluation of TRMM 3B42 precipitation estimates and WRF retrospective precipitation simulation over the Pacific–Andean region of Ecuador and Peru. *Hydrology and Earth System Sciences*, 18(8):3179–3193.

Osorio, J. A. (2007). *El río Tunjuelo en la historia de Bogotá, 1900-1990*. Alcaldia Mayor de Bogotá, Bogotá, primera ed edition.

Pacific Disaster Center (2006). *Bogotá , Colombia Disaster Risk Management Profile*. 3CD City Profiles Series, Bogotá, Colombia.

Papathoma-Köhle, M., Kappes, M., Keiler, M., and Glade, T. (2011). Physical vulnerability assessment for alpine hazards: State of the art and future needs. *Natural Hazards*, 58(2):645–680.

Papathoma-Köhle, M., Keiler, M., Totschnig, R., and Glade, T. (2012). Improvement of vulnerability curves using data from extreme events: debris flow event in South Tyrol. *Natural Hazards*, 64(3):2083–2105.

Patton, P. (1988). Drainage basin morphometry and floods. In Baker, V. R., Kochel, R. C., and Patton, P. C., editors, *Flood Geomorphology*, chapter 3, pages 51–64. Wiley, New York.

Patton, P. and Baker, V. (1976). Morphometry and floods in small drainage basins subject to diverse hydrogeomorphic controls. *Water Resources Research*, 12(5):941–952.

Pebesma, E. J. and Wesseling, C. G. (1998). Gstat: a program for geostatistical modelling, prediction and simulation. *Computers and Geosciences*, 24(1):17–31.

Pérez-Peña, J. V., Azañón, J. M., and Azor, A. (2009). CalHypso: An ArcGIS extension to calculate hypsometric curves and their statistical moments. Applications to drainage basin analysis in SE Spain. *Computers & Geosciences*, 35(6):1214–1223.

Pfannerstill, M., Guse, B., and Fohrer, N. (2014). Smart low flow signature metrics for an improved overall performance evaluation of hydrological models. *Journal of Hydrology*, 510:447–458.

Pokhrel, P., Yilmaz, K. K., and Gupta, H. V. (2012). Multiple-criteria calibration of a distributed watershed model using spatial regularization and response signatures. *Journal of Hydrology*, 418-419:49–60.

Portalés, C., Boronat-Zarceño, N., Pardo-Pascual, J. E., and Balaguer-Beser, A. (2008). Un nuevo método para el cálculo de precipitaciones medias mediante técnicas de interpolación geoestadística considerando las características geográficas y topográficas del territorio. In *IX Congreso Internacional de Ingeniería Geomática y Topográfica*, Valencia.

Prochaska, a., Santi, P., Higgins, J., and Cannon, S. (2008). Debris-flow runout predictions based on the average channel slope (ACS). *Engineering Geology*, 98(1-2):29–40.

Prudhomme, C. (1999). Mapping a statistic of extreme rainfall in a mountainous region. *Physics and Chemistry of the Earth, Part B: Hydrology, Oceans and Atmosphere*, 24(1-2):79 – 84.

Puricelli, M. (2008). *Estimación y distribución de parámetros del suelo para la modelación hidrólogica*. PhD thesis, Universidad Politécnica de Valencia.

Quan Luna, B., Blahut, J., Van Westen, C. J., Sterlacchini, S., Van Asch, T. W. J., and Akbas, S. O. (2011). The application of numerical debris flow modelling for the generation of physical vulnerability curves. *Natural Hazards and Earth System Science*, 11(7):2047–2060.

R Development Core Team (2010). A language and environment for statistical computing. Technical report, R Foundation for Statistical Computing, Vienna, Austria.

Rama Rao, Y. V., Alves, L., Seulall, B., Mitchell, Z., Samaroo, K., and Cummings, G. (2012). Evaluation of the weather research and forecasting (WRF) model over Guyana. *Natural Hazards*, 61(3):1243–1261.

Remesan, R., Bellerby, T., Holman, I., and Frostick, L. (2014). WRF model sensitivity to choice of parameterization: a study of the 'York Flood 1999'. *Theoretical and Applied Climatology*, pages 229–247.

Renard, B., Kavetski, D., Kuczera, G., Thyer, M., and Franks, S. W. (2010). Understanding predictive uncertainty in hydrologic modeling: The challenge of identifying input and structural errors. *Water Resources Research*, 46:1–22.

Rene, J.-r., Madsen, H., and Mark, O. (2012). Probabilistic forecasting for urban water management : A case study. In *9th International Conference on Urban Drainage Modelling*, volume 3, pages 1–11, Belgrade, Serbia. IAHR.

Rengifo, O. (2012). Modelaje de flujos de detritos potenciales a partir de un modelo de elevación digital SRMT (Shuttle Radar Topography Mission): cuenca alta del río Chama , noroeste de Venezuela. *Revista Geográfica Venezolana*, 53(1):93–108.

Reusser, D. E. (2010). *Combining smart model diagnostics and effective data collection for snow catchments*. Dissertation, University of Postdam.

Reyes, O. (2014). Utilización de modelos hidrológicos para la determinación de cuencas en ecosistemas de páramo. *Revista Ambiental Agua, Aire y Suelo*, pages 56–65.

Rickenmann, D. (1999). Empirical Relationships for Debris Flows. *Natural Hazards*, 19:47–77.

Rickenmann, D. and Zimmermann, M. (1993). The 1987 debris flows in Switzerland: documentation and analysis. *Geomorphology*, 8(2-3):175–189.

Rigon, R., Rodriguez-Iturbe, I., and Maritan, A. (1996). On Hack's law. *Water Resources*, 32(11):3367–3374.

Roberts, N. M., Cole, S. J., Forbes, R. M., Moore, R. J., and Boswellc, D. (2009). Use of high-resolution NWP rainfall and river flow forecasts for advance warning of the Carlisle flood, north-west England. *Meteorological Applications*, 16(1):23–34.

Robertson, D. E., Shrestha, D. L., and Wang, Q. J. (2013). Post-processing rainfall forecasts from numerical weather prediction models for short-term streamflow forecasting. *Hydrology and Earth System Sciences*, 17:3587–3603.

Rogelis, M. C. (2006). Sistema de alerta temprana del río Tunjuelo 2006. Technical report, Fondo de Prevención y Atención de Emergencias, Bogotá.

Rogelis, M. C. and Werner, M. (2013). Regional flood susceptibility analysis in mountainous areas through the use of morphometric and land cover indicators. *Nat. Hazards Earth Syst. Sci. Discuss.*, 1(2004):7549–7593.

Rogelis, M. C. and Werner, M. G. F. (2012). Spatial Interpolation for Real-Time Rainfall Field Estimation in Areas with Complex Topography. *Journal of Hydrometeorology*, 14(1):85–104.

Rossa, A., Liechti, K., Zappa, M., Bruen, M., Germann, U., Haase, G., Keil, C., and Krahe, P. (2011). The COST 731 Action: A review on uncertainty propagation in advanced hydro-meteorological forecast systems. *Atmospheric Research*, 100(2-3):150–167.

Rowbotham, D., De Scally, F., and Louis, J. (2005). The identification of debris torrent basins using morphometric measures derived within a GIS. *Geografiska Annaler*, 87(4):527–537.

Ruiz, J. F. (2010). ¿Cómo interpretar los modelos de pronóstico del estado del tiempo? Technical report, Instituto de Hidrología, Meteorología y Estudios Ambientales, Bogotá.

Ruiz-Pérez, M. and Gelabert Grimalt, M. (2012). Análisis de la vulnerabilidad social frente a desastres naturales: el caso de la isla de Mallorca. *GeoSig*, 4:1–26.

Rygel, L., O'Sullivan, D., and Yarnal, B. (2006). A method for constructing a social vulnerability index: An application to hurricane storm surges in a developed country. *Mitigation and Adaptation Strategies for Global Change*, 11(3):741–764.

Saczuk, E. A. R. (1998). *GIS-Based Modeling of Debris Hows in Banff National Park, Alberta*. PhD thesis, University of Manitoba, Winnigpegm Manitoba.

Safaripour, M., Monavari, M., and Zare, M. (2012). Flood Risk Assessment Using GIS (Case Study : Golestan Province , Iran). *Polish Journal of Envirommental Studies*, 21(6):1817–1824.

Sakals, M. E., Innes, J. L., Wilford, D. J., Sidle, R. C., and Grant, G. E. (2006). The role of forests in reducing hydrogeomorphic hazards. *Forest Snow and Landscape Research*, 80(1):11–22.

Salvetti, A., Antonucci, A., and Zaffalon, M. (2008). Spatially Distributed Identification of Debris Flow Source Areas by Credal Networks. In Sànchez-Marrè, M., Béjar, J., Comas, J., Rizzoli, A. E., and Guariso, G., editors, *International Congress on Environmental Modelling and Software (iEMSs 2008)*, pages 380–387, Barcelona, Catalonia. International Environmental Modelling and Software Society.

Santangelo, N., Daunis-i Estadella, J., di Crescenzo, G., di Donato, V., Faillace, P. I., Martín-Fernández, J. A., Romano, P., Santo, A., and Scorpio, V. (2012). Topographic predictors of susceptibility to alluvial fan flooding, Southern Apennines. *Earth Surface Processes and Landforms*, 37(8):803–817.

Santo, A., Santangelo, N., di Crescenzo, G., Scorpio, V., de Falco, M., and Chirico, G. B. (2015). Flash flood occurrence and magnitude assessment in an alluvial fan context: The October 2011 event in the Southern Apennines. *Natural Hazards*, 78(1):417–442.

Santos, R. and Duarte, M. (2006). Topographic signature of debris flow dominated channels implications for hazard assessment. In Lorenzini, G., editor, *International conference on monitoring, simulation, prevention and remediation of dense and debris flows*, pages 301–310, Southampton. WIT Press.

Saulnier, G. M., Obled, C., and Beven, K. (1997). Analytical compensation between DTM grid resolution and effective values of saturated hydraulic conductivity within the TOPMODEL framework. *Hydrological Processes*, 11(9):1331–1346.

Sawicz, K., Wagener, T., Sivapalan, M., Troch, P. a., and Carrillo, G. (2011). Catchment classification: empirical analysis of hydrologic similarity based on catchment function in the eastern USA. *Hydrology and Earth System Sciences*, 15(9):2895–2911.

Scardovi, E. (2015). *Rainfall spatial predictions: a two-part model and its assessment*. PhD thesis, University of Bologna.

Schanze, J., Zeman, E., and Marsalek, J. (2006). *Flood risk management: hazards, vulnerability and mitigation measures*. Springer Netherlands, Dordrecht, The Netherlands.

Schmidtlein, M. C., Deutsch, R., Piegorsch, W. W., and Cutter, S. L. (2008). Building indexes of Vulnerability : a sensitivity analysis of the Social Vulnerability Index. *Risk Analysis*, 28(4):1099–1114.

Schueler, T. (1995). *Site planning for urban stream protection*. Metropolitan Washington Council of Governments. Accessed: 2013-08-25.

Schuurmans, J. M., Bierkens, M. F. P., Pebesma, E. J., and Uijlenhoet, R. (2007). Automatic prediction of high-resolution daily rainfall fields for multiple extents: The potential of operational radar. *Journal of Hydrometeorology*, 8(6):1204 – 1224.

Seethapathi, P., Dutta, D., and Kumar, R. (2008). *Hydrology of small watersheds*. The Energy and Resources Institute, New Deli.

Seidl, M. and Dietrich, W. (1993). The problem of channel erosion into bedrock. *Catena supplement*, pages 101–124. Accessed: 2013-07-10.

Seifert, I., Thieken, A. H., Merz, M., Borst, D., and Werner, U. (2009). Estimation of industrial and commercial asset values for hazard risk assessment. *Natural Hazards*, 52(2):453–479.

Sene, K. (2008). *Flood Warning, Forecasting and Emergency Response*. Springer.

Sene, K. (2013). *Flash Floods: Forecasting and Warning*. Springer Netherlands.

Sevink, J. (2007). Páramo Andino Project. In *Hydrology workshop in Mérida*, page 51, Mérida, Venezuela. PPA project.

Shreve, R. L. (1974). Variation of mainstream length with basin area in river networks. *Water Resources Research*, 10(6):1167–1177.

Sodnik, J. and Miko, M. (2006). Estimation of magnitudes of debris flows in selected torrential watersheds in Slovenia. *Acta geographica Slovenica*, 46(1):93–123.

Sterlacchini, S., Akbas, S. O., Blahut, J., Mavrouli, O.-C., Garcia, C., Luna, B. Q., and Corominas, J. (2014). Methods for the characterization of the vulnerability of elements at risk. In Van Asch, T., Corominas, J., Greiving, S., Malet, J.-P., and Sterlacchini, S., editors, *Mountain Risks: From Prediction to Management and Governance*, volume 34, pages 233–273. Springer Netherlands, Dordrecht, The Netherlands.

Stock, J. and Dietrich, W. E. (2003). Valley incision by debris flows: Evidence of a topographic signature. *Water Resources Research*, 39(4):1089.

Su, M. and Kang, J. (2005). A grid-based GIS approach to regional flood damage assessment. *Journal of Marine Science and Technology*, 13(3):184–192.

Syed, K. H., Goodrich, D. C., Myers, D. E., and Sorooshian, S. (2003). Spatial characteristics of thunderstorm rainfall fields and their relation to runoff. *Journal of Hydrology*, 271(1-4):1–21. doi: DOI: 10.1016/S0022-1694(02)00311-6.

Tabios, G. Q. and Salas, J. D. (1985). A comparative analysis of techniques for spatial interpolation of precipitation. *Journal of the American Water Resources Association*, 21(3):365–380.

Teng, J., Potter, N. J., Chiew, F. H. S., Zhang, L., Wang, B., Vaze, J., and Evans, J. P. (2015). How does bias correction of RCM precipitation affect modelled runoff ? *Hydrology and Earth System Sciences*, 19(2):711–728.

Theis, S. E., Hense, A., and Damrath, U. (2005). Probabilistic precipitation forecasts from a deterministic model: a pragmatic approach. *Meteorological Applications*, 12:257.

Thieken, A., Merz, B., Kreibich, H., and Apel, H. (2006). Methods for flood risk assessment: concepts and challenges. In *International Workshop on Flash Floods in Urban Areas*, pages 1–12, Muscat – Sultanate of Oman.

Thiemig, V. (2014). *The development of pan-African flood forecasting and the exploration of satellite-based precipitation estimates*. PhD thesis, Utrecht University.

Tobón, C. and Gil - Morales, E. G. (2007). Capacidad de interceptación de la niebla por la vegetación de los páramos andinos. *Avances en Recursos Hidraulicos*, 15(1):35–46.

Todini, E., Alberoni, P., and Butts, M. (2005). Understanding and reducing uncertainty in flood forecasting. In *International Conference on Innovation Advances and Implementation of Flood Forecasting Technology*, pages 1–43.

Torres, V., Vandenberghe, J., and Hooghiemstra, H. (2005). An environmental reconstruction of the sediment infill of the Bogotá basin (Colombia) during the last 3 million years from abiotic and biotic proxies. *Palaeogeography, Palaeoclimatology, Palaeoecology*, 226(1-2):127–148.

Totschnig, R. and Fuchs, S. (2013). Mountain torrents: Quantifying vulnerability and assessing uncertainties. *Engineering Geology*, 155:31–44.

Totschnig, R., Sedlacek, W., and Fuchs, S. (2011). A quantitative vulnerability function for fluvial sediment transport. *Natural Hazards*, 58(2):681–703.

Tsao, T.-c., Hsu, W.-k., Cheng, C.-t., Lo, W.-c., and Chen, C.-y. (2010). A preliminary study of debris flow risk estimation and management in Taiwan. *International symposium interpraevent in the Pacific Rim—Taipei*, pages 930–939.

Tucker, G. and Bras, R. (1998). Hillslope processes, drainage density, and landscape morphology. *Water Resources Research*, 34(10):2751–2764.

UNAL and INGEOMINAS (2007). Generación de una guía metodológica para la evaluación de la amenaza por movimientos en masa tipo flujo, caso piloto Cuenca Quebrada La Negra. Technical report, INGEOMINAS, Bogotá.

UNEP (2003). *Assessing Human Vulnerability to Environmental Change: Concepts, Issues, Methods and Case Studies*. United Nations Environment Programme, Nairobi, Kenya.

UNISDR (2009). Terminology on Disaster Risk Reduction. Technical report, United Nations International Strategy for Disaster Reduction, Geneva, Switzerland.

United Nations and ISDR (2004). *Living with risk: a global review of disaster reduction initiatives*, volume 1. United Nations Publications, Geneva, Switzerland.

United Nations General Assembly (2015). Resolution adopted by the General Assembly on 3 June 2015. Sendai Framework for Disaster Risk Reduction 2015–2030.

USACE (2000). Hydrologic Modeling System HEC-HMS. Technical report, U.S. Army Corps, Davis.

Van de Beek, C. Z., Leijnse, H., Torfs, P., and Uijlenhoet, R. (2011). Climatology of daily rainfall semivariance in the netherlands. *Hydrol. Earth Syst. Sci.*, 15(1):171–183.

Van Westen, C., Kappes, M., Luna, B., Frigerio, S., Glade, T., and Malet, J.-P. (2014). Medium-Scale Multi-hazard Risk Assessment of Gravitational Processes. In Van Asch, T., Corominas, J., Greiving, S., Malet, J.-P., and Sterlacchini, S., editors, *Mountain Risks: From Prediction to Management and Governance*, pages 201–231. Springer Netherlands, Dordrecht, The Netherlands.

Varnes, D. J. (1984). Landslide hazard zonation: a review of pronciples and practice. Technical report, Commission on landslides of the IAEG, UNESCO.

Vergara, W., Deeb, A., and Leino, I. (2011). Assessment of the impacts of climate change on mountain hydrology: development of a methodology through a case study in the Andes of Peru. Technical report, The World Bank, Washington, D.C.

Verkade, J. S., Brown, J. D., Reggiani, P., and Weerts, a. H. (2013). Post-processing ECMWF precipitation and temperature ensemble reforecasts for operational hydrologic forecasting at various spatial scales. *Journal of Hydrology*, 501:73–91.

Verkade, J. S. and Werner, M. G. F. (2011). Estimating the benefits of single value and probability forecasting for flood warning. *Hydrology and Earth System Sciences*, 15(12):3751–3765.

Vidal, I. and Varas, E. (1982). Estimaciones de precipitacion en lugares con poca informacion. *Agricultura Técnica*, 42(1):23 – 30.

Vincendon, B., Ducrocq, V., Nuissier, O., and Vié, B. (2011). Perturbation of convection-permitting NWP forecasts for flash-flood ensemble forecasting. *Natural Hazards and Earth System Science*, 11(5):1529–1544.

Wan, S., Lei, T., Huang, P., and Chou, T. (2008). The knowledge rules of debris flow event: A case study for investigation Chen Yu Lan River, Taiwan. *Engineering Geology*, 98(3-4):102–114.

Weissmann, G. S., Bennett, G. L., and Lansdale, a. L. (2005). Factors controlling sequence development on Quaternary fluvial fans, San Joaquin Basin, California, USA. *Geological Society, London, Special Publications*, 251(1):169–186.

Welsh, A. J. (2007). *Delineating debris-flow hazards on alluvial fans in the Coromandel and Kaimai regions, New Zealand, using GIS*. PhD thesis, University of Canterbury, Canterbury.

Welsh, A. J. and Davies, T. (2010). Identification of alluvial fans susceptible to debris-flow hazards. *Landslides*, 8(2):183–194.

Wilford, D. J., Sakals, M. E., Innes, J. L., Sidle, R. C., and Bergerud, W. a. (2004). Recognition of debris flow, debris flood and flood hazard through watershed morphometrics. *Landslides*, 1(1):61–66.

Wilks, D. S. (2006). *Statistical methods in the atmospheric sciences*, volume 91. Academic Press, San Diego, California, USA, second edition.

Willemin, J. H. (2000). Hack ' s law: Sinuosity , convexity , elongation. *Water Resources Research*, 36(11):3365–3374.

Willgoose, G. and Hancock, G. (1998). Revisiting the hypsometric curve as an indicator of form and process in transport-limited catchment. *Earth Surface Processes and Landforms*, 23(7):611–623.

Wisner, B., Blaikie, P., Cannon, T., and Davis, I. (2003). *At Risk: natural hazards, people's vulnerability and disasters*. Routledge, London, UK, second edition.

Wu, S., Li, J., and Huang, G. H. (2007). Modeling the effects of elevation data resolution on the performance of topography-based watershed runoff simulation. *Environmental Modelling and Software*, 22:1250–1260.

Yang, W., Andréasson, J., Phil Graham, L., Olsson, J., Rosberg, J., and Wetterhall, F. (2010). Distribution-based scaling to improve usability of regional climate model projections for hydrological climate change impacts studies. *Hydrology Research*, 41(3-4):211.

Yilmaz, K. K., Gupta, H. V., and Wagener, T. (2008). A process-based diagnostic approach to model evaluation: Application to the NWS distributed hydrologic model. *Water Resources Research*, 44(9).

Yuan, X. and Wood, E. F. (2012). Downscaling precipitation or bias-correcting streamflow? Some implications for coupled general circulation model (CGCM)-based ensemble seasonal hydrologic forecast. *Water Resources Research*, 48:1–7.

Yucel, I., Onen, a., Yilmaz, K. K., and Gochis, D. J. (2015). Calibration and evaluation of a flood forecasting system : Utility of numerical weather prediction model , data assimilation and satellite-based rainfall. *Journal of Hydrology*, 523:49–66.

Zavoianu, I. (1985). *Morphometry of drainage basins / Ion Zavoianu;[translated from the Romanian by Adriana Ionescu-Parau]*. Elsevier; Editura Academiei Republicii Socialiste Romania, Amsterdam.

Zhang, X., Hörmann, G., Gao, J., and Fohrer, N. (2011). Structural uncertainty assessment in a discharge simulation model. *Hydrological Sciences Journal*, 56(February 2015):854–869.

Curriculum Vitae

María Carolina Rogelis was born in Bogotá, Colombia in 1977. She studied Civil Engineering at the Universidad Nacional de Colombia, where she graduated in 1999. From then until 2001 she studied a Master of Engineering in Water Resources management in Los Andes University in Bogotá, while working part time for the Empresa de Acueducto y Alcantarillado de Bogotá.

From 2002 to 2004 she studied at UNESCO-IHE in Delft, The Netherlands, where she obtained an MSc degree in Hydraulic Engineering with distiction. In 2004 she joined the Fondo de Prevención y Atención de Emergencias de Bogotá as flood risk management specialist, contributing in the development of flood early warning systems and flood risk assessement for Bogotá. In 2009, she started her PhD research at UNESCO-IHE, undertaking a program partly based in The Netherlands and partly in Bogotá, focused on operational flood forecasting, warning and response for multi-scale flood risk in developing cities.

Mrs. Rogelis worked during 7 years in flood management and flood forecasting systems at the Fondo de Prevención y Atención de Emergencias de Bogotá, and from 2011 she works as a consultant in the field of flood risk management for The World Bank and Colombian companies.

T - #0406 - 101024 - C48 - 244/170/12 - PB - 9781138030039 - Gloss Lamination